D1586175

Biomedical Calculations

Principles and Practice

Richard F. Burton
Institute of Biomedical and Life Sciences
University of Glasgow, UK

John Wiley & Sons, Ltd

Other Wiley Editorial Offices

John Wiley & Sons Inc., 111 River Street, Hoboken, NJ 07030, USA

Jossey-Bass, 989 Market Street, San Francisco, CA 94103-1741, USA

Wiley-VCH Verlag GmbH, Boschstr. 12, D-69469 Weinheim, Germany

John Wiley & Sons Australia Ltd, 33 Park Road, Milton, Queensland 4064, Australia

John Wiley & Sons (Asia) Pte Ltd, 2 Clementi Loop #02-01, Jin Xing Distripark, Singapore 129809

John Wiley & Sons Canada Ltd, 6045 Freemont Blvd, Mississauga, Ontario, L5R 4J3

Wiley also publishes its books in a variety of electronic formats. Some content that appears in print may
not be available in electronic books.

Library of Congress Cataloging-in-Publication Data

Burton, R. F. (Richard F.)
    Biomedical calculations : principles and practice / Richard F. Burton.
      p.   cm.
    Includes bibliographical references and index.
    ISBN 978-0-470-51910-3 (cloth) – ISBN 978-0-470-51911-0 (pbk.)
1. Medical sciences–Mathematics.   I. Title.
    R853.M3B875 2008
    610.1′53–dc22
                                                                                2008016830

British Library Cataloguing in Publication Data

A catalogue record for this book is available from the British Library

ISBN 978-0-470-51910-3 (HB)   978-0-470-51911-0 (PB)

Typeset in 10.5/13pt Minion by Aptara Inc., New Delhi, India
Printed and bound in Singapore by Markono Print Media Pte Ltd
This book is printed on acid-free paper.

Contents

Preface

This is a guide to quantitative thinking in the biomedical sciences for students and professionals. Many students are unsure about such basic concepts as 10^{-3}, $x^{1/2}$, $\log x$, reciprocal, percentage, microlitre, millimoles per litre, order of magnitude and calories. Other people who understand these well may yet falter in applying them – whether to straightforward situations like preparing drug solutions or to more interesting quantitative problems. The book is intended to help all such people. Its backbone is one simple, but neglected idea that can transform one's abilities. However, because no such idea can suffice if one lacks basic concepts like those I have just listed, I try to provide these in a way that will not alienate those who already have them. The biomedical examples may prove interesting in themselves, for some are not to be found in other textbooks.

I do not write books without some personal inner drive. (Moreover, echoing Spike Milligan, 'I vowed never to write another [science] book – and this is it'.) I am not innately clever at solving quantitative problems myself, but a few years ago I came upon the simple key, or guiding principle, to what I like to call 'calculating science'. I do not recall when the penny dropped, but there is nothing like revelations late in life for creating zealots! What I realized was that, if I paid more attention to the units involved (e.g. grams per litre, kg m s^{-2} etc.), answers to calculations could sometimes emerge automatically with little need for further reasoning – and my mistakes were fewer. The benefits were most obvious when a weary brain was applied to unfamiliar aspects of biology. Some readers will think I discovered the obvious. Indeed page four of the physics textbook I used at school implies in one paragraph much of what struck me only later as a revelation. However, that otherwise excellent book gave the matter too little emphasis and did not include units in the working of its subsequent calculations. Many recent physics textbooks also fail to do so. So what is really an ancient principle is too often neglected. Regardless of what schoolbooks may say about it and teachers teach, there are certainly many university students, and indeed professional scientists, who have not acquired this key to confident quantitative thinking.

A good name for this idea is 'unit analysis'. I spell it out in Chapter 1 and hope that its full nature and power will be further revealed throughout the book. So far I am emphasizing the value of unit analysis in calculations, but it may also be applied in contexts where no numbers are involved, as when one is trying to make sense of formulae and equations. In that context, readers familiar with 'dimensional analysis' will see that the two ideas are closely related. Dimensional analysis, which is outlined in Chapter 1, can be more elegant, but is often less helpful in biological contexts. For consistency it is just unit analysis that I use here.

My first proper research, which was about the shapes of leaves and long ago, would have proceeded faster if I had understood dimensional or unit analysis. Now the challenge of writing this book has led me into a sideline of research (anthropometry) that I had no reason to contemplate otherwise. Either technique of analysis can prove invaluable as one tries to understand relationships amongst biological variables, and anthropometry has given me some useful illustrations of this.

Unit analysis calls for an understanding of units and, following a unit theme, I look successively at length, area, volume, mass, moles and equivalents, combining some of these as units for density and concentration. Time is added later, allowing treatment of speed, acceleration and rates of flow. Subsequent chapters add force and pressure, and then energy, work, power and temperature. Other topics are integrated into this progression according to the following guiding principles.

- The mathematics should be simple, involving nothing more advanced than exponents, logarithms and very basic algebra, and help is offered even with these. Topics requiring logarithms are confined to the final chapters.

- Formulae are not just to be applied, but, where possible, to be understood.

- Plotting and interpreting graphs are necessary skills.

- There are topics of enough general importance that they need to be included, such as dosage calculation, renal clearance, gas mixtures, buffering and acid–base balance. Not every reader will require these, but I try to make more general points in dealing with them. As I like to illustrate, a particular quantitative approach – a way of thinking or form of calculation – may apply in varied contexts.

- Not all the biological topics need to be as exhaustively treated here as they are in standard courses and textbooks – so long as sufficient background is provided.

- Desirable as it may be to work with a consistent set of units (e.g. SI), one should be able to cope happily with the variety of units that are inevitably encountered in the real world. Consistency is therefore unhelpful in this book.

- The reader should be offered problems for practice, preferably problems having useful or informative answers.

There is no one ideal way of selecting and ordering my topics and of balancing levels of difficulty for a diverse readership, but I trust that my choices prove helpful and interesting.

Acknowledgements

I thank Dr Dorothy Aidulis for her comments and Dr Francis Burton for reading much of the manuscript and helping in the preparation of figures.

Richard F. Burton

This book, and how to use it

A major theme is unit analysis, for this can facilitate calculations and help one to understand, and sometimes even improve, equations and formulae. What I mean by unit analysis is explained in Part I. Physical units are important to the organization of the book.

Part II involves mainly units of length, area, volume and mass used singly and in combination, but also moles, equivalents and osmoles. These units are used in treating density, concentrations and drug dosages.

Part III introduces no new units, being largely concerned with areas, volumes and masses in relation to erythrocytes and the human form. It also discusses graphs and the different kinds of formulae and equations that one meets in biomedical science.

In Part IV the introduction of the time element allows discussion of rates and how to think about particular formulae that involve them.

Part V adds units of force, pressure, energy, work, power and temperature. Pressure and temperature play a vital role in the concluding chapter of this section, which is on respiratory gases.

Part VI introduces logarithms, opening the way for the Nernst and Henderson–Hasselbalch equations, the graphical analysis of exponential time courses, dose–response curves, buffering and acid–base balance.

The book thus has a developing sequence and may be read accordingly. Nevertheless, because the sequence is not much based on any standard divisions of biomedical science, readers will no doubt read selectively according to their own current needs and interests. Remember, however, that the book is mainly about coping with calculations and quantitative thinking, and that there are lessons to be learnt from topics that might otherwise seem personally irrelevant,

Numerical problems are provided and answers, with working, are given in Appendix D. Most of the problems are gathered at the ends of chapters, but some are presented individually in boxes where they are particularly apposite in the main text.

Calculations should be written out in full, with units included throughout the working; correct numerical answers are usually less important than how they are obtained.

Most textbook writers aim for consistency in their choice and abbreviation of units. I do not, but try, instead, to familiarize the reader with the variety and inconsistencies of units in the 'real world'. Words and unit symbols are sometimes mingled deliberately in ways that contravene the rules of scientific style, but which can make thinking easier.

Little mathematical knowledge is assumed beyond the basic operations $(+, -, \times, \div, \sqrt{}, \text{indices})$. Nevertheless notes on mathematical language and symbols are available in Appendix A. Other notes are provided in Appendix C.

Part I

1

Unit analysis: the neglected key to confidence

The Preface emphasizes the importance I attach to unit analysis. This chapter explains more about what it means and shows how it can aid thinking in a variety of circumstances.

1.1 Calculating with units

Most calculations in biology and biomedical science involve units and these should be included in the calculations along with the numbers. Numbers unaccompanied by units may be called abstract, pure or dimensionless. At the risk of being embarassingly elementary, we start with three of these, all integers.

Given the following arithmetical calculation, the easiest way to a solution is to cancel the fours (which is just a matter of recognizing that 4 divided by 4 equals 1):

$$\frac{3}{4} \times \frac{4}{2} = \frac{3}{2} = 1.5.$$

The numbers 2, 3 and 4 may be replaced by symbols such as x, y and z, representing unknown or unspecified numbers to produce an algebraic expression:

$$\frac{x}{y} \times \frac{y}{z}.$$

Biomedical Calculations   Richard F. Burton
© 2008 John Wiley & Sons, Ltd

Since the same cancellation process is possible (because $y/y = 1$), the answer is x/z. The point now is that the same algebraic treatment can be applied to units. As a specific example, one might have to multiply, say, 2 grams per litre by 9 litres per minute. In doing so, one may write:

$$2\frac{\text{gram}}{\text{litre}} \times 9\frac{\text{litre}}{\text{minute}} = 18\frac{\text{gram}}{\text{minute}} = 18\,\text{gram/minute}.$$

The numbers are multiplied in the usual way and, as for the units, we have 'litre' divided by 'litre' – which equals 1. It usually helps to put a line through each of the terms or units that are being cancelled:

$$\frac{\text{gram}}{\cancel{\text{litre}}} \times \frac{\cancel{\text{litre}}}{\text{minute}} = \text{gram/minute}.$$

This particular calculation could arise in solving a problem relating to the rate at which some unspecified substance, at a concentration of 2 grams per litre, is being carried along in fluid that is flowing at a rate of 9 litres per minute.

   Here is a similar problem, again deliberately simple. Cars are travelling along a road in one direction at a rate of 10 per minute. There are two people in each car. At what rate are people travelling along the road? One can write the calculation as above, cancelling the 'cars':

$$\text{Rate} = 2\frac{\text{person}}{\text{car}} \times 10\frac{\text{car}}{\text{minute}} = 20\,\text{person/minute}.$$

This may be seen as an even simpler problem because the situation is mundane and easy to relate to, but the reasoning is no different. In any case, it is so easy to solve that few people would bother to spell out the details of the calculation, especially the units (i.e. 'person', 'car'). Indeed some students are reluctant to write out units, even with harder problems and when they are specifically instructed to do so. Too often they then get the answer wrong.

   Readers are therefore urged to write out all calculations in full at all stages, including units.

   One generally uses symbols for units, for example, g and kg for gram and kilogram, min for minute, s for second and m for metre. I use L for litre, rather than l, because the latter can look like 'one' or capital I both with handwriting and with many typefaces. Many journals require the use of 'L'.

1.2 Ways of writing composite units

I have been expressing composite units (gram/litre, gram/minute, cars/minute, etc.) using the slash or solidus ('/') to mean 'per'. In that context it simultaneously also means 'divided by'. Sometimes, especially in older writings, the slash may be used twice within a single set of units – as, for example, m/s/s for an acceleration. This is to be read as 'metres per second per second' with the slash then standing only ambiguously for 'divided by'. There is clearly the possibility for confusion with this usage and a more proper representation of the units nowadays is m/s^2. Some people write 'm/s.s', but that could be read as m/s × s, which equals m. Moreover, SI symbols like m and s should not be followed by full stops (periods) except at the ends of sentences. The slash can be read unambiguously as standing for division, so that another manifestation of the same combination of units is m s^{-2}. Some people assert that it is now more usual and proper to express units using indices where appropriate, with negative values where a slash would otherwise be used. To ensure familiarity with both systems, both are used in this book, but with the slash often favoured where the act of division is being emphasized. Indices can be clearer when three or more units are combined, as in kg m^2 s^{-2}. Indices are explained in Appendix A, along with other aspects of mathematical notation.

1.3 How unit analysis can guide thinking and help solve problems

I have asserted that attention to units can sometimes cause answers to emerge automatically and with little need for further reasoning. I give other examples later, but for the moment here is a very simple one. A difficulty in choosing a good illustration is that, if the problem is too easy, unit analysis provides no apparent benefit. Here I cheat, therefore, by giving you the necessary data, but not telling you the question. That way you cannot find the answer by the more usual processes of reasoning. So here is the problem.

   You are given the following two items of data, both of which, and no others, are needed to solve the problem:

2 gram NaCl

0.1 litre water

What is the answer, and what is the most likely question?

   Assuming that the two quantities are to be combined arithmetically in a simple manner, one might try adding them, or subtracting one from the other. However the units are different, and *it is inadmissible to add or subtract quantities with different units*. To do so would be as meaningless (mathematically) as adding, say, three cats and two bicycles. It is permissible to multiply the two original quantities together, but the product, namely 0.2 (g NaCl)×(L water), has no obvious meaning. The remaining simple possibility is to divide one by the other, to obtain either 20 g NaCl/L water or 0.05 L water/g NaCl. One might find a context for the second of these quantities, but it is the first that has the most obvious meaning: it is a concentration of salt in water. The question was therefore probably 'What is the concentration of NaCl when 2 g of it are dissolved in 0.1 L of water?'

One can hardly solve every problem this way. If the two items of data above had had the same units, these being, say, 5 m and 7 m, then addition, subtraction, multiplication and division would each produce meaningful quantities, one of these being an area (35 m^2).

1.4 When to specify substances along with units

Note that the answer to that unstated question is written above as 20 g NaCl/L water rather than as 20 g/L (or 20 g L^{-1}). When should substances be specified with units? Suppose that 20 mL alcohol is diluted with 50 mL of water. The resulting concentration is 0.4 mL alcohol/mL water. If the latter is expressed just as 0.4 mL/mL, the meaning could be uncertain. For example, it could be interpreted as 0.4 mL alcohol per mL of mixture. One could go further and simplify 0.4 mL/mL to just 0.4, cancelling the identical volume units; the result would be even more ambiguous, for then there would be no indication of the relevant measure of quantity (mL, g, mole, etc.). Especially when indices are used, the use of brackets can make the units clearer, e.g. (mL alcohol) (mL water)$^{-1}$.

Some writers on scientific style disapprove of this mingling of words and units, but that is in the context of formal writing, as for theses and journal articles. My concern here is with correct and easy calculation, and what we write on the back of an envelope is our own business.

Let us consider a more physiological example now. The concentration of oxygen in arterial blood is typically about 0.2 L O_2/L blood. Suppose that one needs to calculate the volume of oxygen in 3 L of blood. To do this, one might write:

$$0.2 \frac{L\, O_2}{(L\, \text{blood})} \times 3 (L\, \text{blood}) = 0.6\, L\, O_2.$$

Here too the brackets, though not actually necessary, are used for clarity, perhaps thereby making '(L blood)' a slightly more obvious term for cancellation. Part of the point of spelling out the units is the avoidance of mistakes. Reassuringly, in this calculation the unit of the answer, i.e. L O_2, corresponds to a volume as required. Suppose, however, that one proceeded with the calculation expressing the data as 0.2 L/L and 3 L (i.e. without specifying what the volumes refer to). Suppose, too, that one calculated wrongly, writing:

$$0.2\frac{L}{L} \div 3\,L = 0.2\frac{L}{L} \times \frac{1}{3\,L} = 0.067\,L^{-1}.$$

One might fail to recognize that the answer is badly wrong numerically, but it should be obvious that 'L^{-1}' cannot be right, because the question asks for a volume, not the reciprocal of a volume.

A few lines above is the expression

$$0.2\frac{L\,O_2}{(L\,blood)},$$

with the comment that

$$0.2\frac{L\,O_2}{L\,blood}$$

is satisfactory too. It is also possible to write

$$\frac{0.2\,L\,O_2}{L\,blood},$$

combining number and units in one expression, and it is this method that is used in this book more often than not. It is frequently this that best conveys the logic of a calculation.

1.5 The need to use appropriate and compatible units in formulae

Physical formulae are generally quoted without specification of units, the unwritten assumption being that all the quantities involved need to be expressed in appropriate and compatible units. 'Everybody knows' that $E = mc^2$, but Einstein's formula is quantitatively untrue if, say, energy (E) is expressed in calories, mass (m) in pounds and the speed of light (c) in miles per hour. There is a general lesson to be learnt from this formula, for we can use it to find an appropriate combination of units for energy. We do this by writing down those for the expression mc^2. If mass is expressed in kilograms and the speed of light is in metres per second, then the combination for mc^2 is:

$$kg \times (m/s)^2 = kg\,m^2/s^2 = kg\,m^2s^{-2}.$$

These are not commonplace units for energy, like calories, joules or British Thermal Units, but they are valid. In fact 1 kg m^2 s^{-2} equals 1 joule by definition. The name joule was coined to avoid the clumsy 'kg m^2 s^{-2}'. Energy units are discussed in Chapter 14.

1.6 Checking and deriving formulae

Unit analysis can be helpful if one needs to use a particular formula and is unsure if one is remembering it correctly. It is also possible almost to derive some formulae for oneself by applying the same principles. Here are two examples. The first is very simple. The second is more complicated to make the point more forcefully.

1. Calling on fading memories, a student might recollect that the volume and surface area of a sphere of radius r are $4\pi r^2$ and $^4/_3\pi r^3$, but without remembering which is which. With r measured in cm, say, the units for r^2 and r^3 are respectively cm^2 and cm^3. These correspond to area and volume, respectively. One therefore concludes that

$$\text{Volume of sphere} = {}^4/_3\pi r^3, \tag{1.1}$$

$$\text{Surface area of a sphere} = 4\pi r^2. \tag{1.2}$$

2. There is a formula relating the period of a simple pendulum to the pendulum length l. The period is the time for one complete swing back and forth. Gravity is relevant, so the formula may be assumed to include g, the acceleration due to gravity. Suspecting that a pendulum would not swing at all in the absence of gravity and would swing faster with increasing gravitational pull, and knowing also that long pendulums have long periods (i.e. swing more slowly), one might try out this formula:

$$\text{period} = A \times \frac{l}{g},$$

where A is a constant of unknown value. Arbitrarily choosing to work with seconds and metres, one spells out the units for both sides of the equation as follows:

$$s = \frac{m}{(m/s^2)} = m \times \frac{s^2}{m} = s^2.$$

Since the formula leads to seconds being equated with seconds squared, it has to be wrong. In terms of units, it can be corrected by introducing a square root:

$$\text{period} = A \times \sqrt{\frac{l}{g}}.$$

(If square roots and the insect's-leg symbol are unfamiliar, see Appendix A.) Unit analysis now gives a satisfactory result:

$$s = \sqrt{\frac{m^2}{m/s}} = \sqrt{\frac{1}{1/s^2}} = \sqrt{(s^2)} = s.$$

In the actual formula, A is replaced by 2π (for reasons that do not concern us here) to give:

$$\text{period} = 2\pi \sqrt{\frac{l}{g}}. \tag{1.3}$$

The unnecessary multiplication sign has been dropped. (For the formula to apply exactly, the angle of swing must be small, and both friction and air resistance must be negligible.)

The pendulum formula is rarely directly relevant to biomedical and life sciences, but it serves its purpose here, and swinging pendulums are easily visualized. Nevertheless there is a link to medical practice. Galileo, who was the first to investigate the pendulum, began his studies in 1583 using his pulse to time the swinging of a cathedral lamp in Pisa. Having observed the regularity of the swing, and the independence of its period from its extent, he constructed a special pendulum of adjustable length for measuring pulse rates at the sick bed.

1.7 When unit analysis raises questions about formulae

Just occasionally unit analysis can help one spot mistakes in formulae, the commonest of these being misprints. However, as will be discussed later, not every useful formula needs to stand up to unit analysis, especially those that are empirical and have no clear and exact physical derivation.

Body fat can be assessed by means of 'skinfold thickness measurements' – measurements of the thickness of folds of skin, with accompanying adipose tissue, at carefully selected sites on the body. Body fat is estimated from these by means of empirical formulae that may include separate terms for age, sex, etc., that do not immediately concern us. The point here is that thickness is measured in millimetres while fat, whether expressed as a percentage (a unit-free number) or in kilograms, is not. Unit analysis suggests both that there is something missing from the formulae and what sort of additional measurement might be included in order to improve them. We come back to this in Chapter 12, which illustrates other ways in which unit analysis can help one to understand aspects of the human body-form.

1.8 Dimensional analysis

In applying Eqs. (1.1)–(1.3) to actual spheres and pendulums, one must specify particular units of distance (i.e. length and linear measurements generally) and of time. Whatever the choice, consistency must be maintained throughout the working. Thus, for the pendulum formula, the acceleration needs to be expressed in the same time and distance units as are used for the period and pendulum length. There is another approach that obviates the need to choose particular units. This is called dimensional analysis.

Instead of analysing formulae and expressions in terms of units, one uses symbols standing for length (distance), time, mass, etc., these three being respectively L, T and M. From these are obtained the dimensions of other quantities, such as velocity ($L\,T^{-1}$), density ($M\,L^{-3}$) and so on. Quantities without units ('dimensionless numbers'), like the 2π in Eq. (1.3) for the pendulum, are ignored. The analysis of Eq. (1.3) is then as follows:

$$T = \sqrt{L} \div \sqrt{\frac{L}{T^2}} = \sqrt{L} \div \frac{\sqrt{L}}{\sqrt{T^2}} = \sqrt{L} \times \frac{T}{\sqrt{L}} = T.$$

To these three dimensions, L, T and M, may be added others, including temperature and electric current. Fuller accounts are to be found in physics textbooks.

Unit analysis has two advantages over dimensional analysis that have already been illustrated. Firstly, it helps one to avoid the inappropriate mixing of units for a given dimension (e.g. seconds with minutes, or litres with cubic centimetres). Secondly, one may more naturally specify, along with physical units, whatever these refer to – e.g. 'L O_2' when these are to be distinguished from 'L blood', as above.

The phrase 'dimensional analysis' is sometimes used in a different sense than this in relation to the calculation of drug dosage. Then the usage corresponds to 'unit analysis' as I use it, but just in the limited context of avoiding errors in the conversion of units. This is discussed in Chapter 8.

Unfortunately 'dimension' has other meanings in common speech and mathematics. When referring to dimensional analysis, one can therefore be completely misunderstood by the uninitiated, so that is another reason for avoiding the word here. 'Dimensionless number' is nevertheless a useful phrase for describing quantities that do not involve units. Percentages (Chapter 3) are dimensionless numbers.

Part II

The units in Part II involve length, area, volume, mass and amount of chemical substance. There is little mention at this stage of concepts such as time, force, pressure, energy and power, for these are mainly deferred to Part IV. However, even the limited range of units treated here is enough to allow general discussion of the Système International (SI) and the practicalities of converting between units, problem solving, making up solutions and calculating drug doses. Other topics include density and the various ways of expressing concentrations. Unit analysis has an important role in all this. Percentages, though unit-free, have been allowed to intrude into the orderly sequence of topics because they cause more problems than our school books warn us about.

Part II

2
Units: length, area, volume, mass, moles and equivalents

2.1 The Système International and unit prefixes

Physics textbooks advocate working within a consistent system known as the Système International, abbreviated to SI. The fundamental units with their symbols include the metre (m), the kilogram (kg), the mole (mol), the second (s) and, for temperature, the Kelvin (K). They may be combined in various ways to produce 'derived' units such as m^2 (for area), $m\ s^{-1}$ (for speed and velocity) and $kg\ m^{-3}$ (for density or concentration).

There is a standard system of prefixes used with SI units for naming units that are, for instance, a thousand times as great (kilo-) or smaller by factors of a thousand or a million (milli- and micro-, respectively). These prefixes, like the base units, have their standard forms. Thus, the sequence for the metre includes the nanometre (nm), micrometre (μm), millimetre (mm), metre (m) and kilometre (km). The sequence extends upwards and downwards in factors of 1000 and 1/1000, respectively. However, supplementing this standard sequence of prefixes we also have, for convenience, 'deci-' and 'centi-' as in decimetre (dm) and centimetre (cm). These prefixes are summarized in Table 2.1. The first three have little biomedical relevance, except at times of major calamity. Some people advocate writing out units in full when there is a danger of hand-written drug prescriptions being mis-read.

Biomedical Calculations   Richard F. Burton
© 2008 John Wiley & Sons, Ltd

Table 2.1 Prefixes for multiples and submultiples of SI units

Multiple	Prefix	Symbol
10^{12}	tera	T
10^{9}	giga	G
10^{6}	mega	M
10^{3}	kilo	k
10^{-1}	deci	d
10^{-2}	centi	c
10^{-3}	milli	m
10^{-6}	micro	μ
10^{-9}	nano	n
10^{-12}	pico	p
10^{-15}	femto	f

Adherence to the SI system is certainly advantageous in most contexts, but for those working in the biomedical and life sciences reality is more complicated. For example, volumes are routinely measured, not in m^3, but in litres (L) and millilitres (mL). In addition, non-SI units like inches and pounds are encountered in older or non-scientific writings. Here we look briefly at units of length, area, volume and mass. Amounts of particular chemical substances may be expressed not only as masses, but also in terms of moles or equivalents, so these too are discussed.

Whether students are most familiar with grams, moles or equivalents in dealing with concentrations may depend partly on which part of the world they are studying in, and where their textbooks come from. Moreover, although this chapter is mainly about SI units, one must be able to cope with others occasionally, such as inches, pounds and gallons. Appendix B is a conversion table relating a number of SI and non-SI units. The general topic of converting between units is discussed at the end of this chapter.

2.2 Length and distance

For length and distance, the SI units of nanometre (nm), micrometre (μm), millimetre (mm), centimetre (cm), decimetre (dm), metre (m) and kilometre (km) have already been mentioned. Americans write 'meter'. For kilometre or kilometer there are two pronunciations nowadays, with the correct one losing out to the newer version in which the accent has moved to the O, as in barómeter. Especially in older works on microscopy and histology, one meets the Ångstrom unit (Å), which equals 0.1 nm. The micrometre (μm) is commonly referred to as a 'micron' or, colloquially, as a 'mu', pronounced as in 'sea mew' and 'emu'. In old histology texts, you may find the micrometer abbreviated just as μ. The nanometer (10 Å) was once called the millimicron (mμ). One sees the value of standardization!

2.3 Area

Areas are quantified in terms of numbers of squares with sides of unit length, which one can call 'square metres', 'square inches', etc. However, the corresponding units are properly written as m^2, in^2, etc. There seem to be no special names in common use for small areas, although for areas of land large enough to plough or fence off we have acres and hectares.

2.4 Volume

Volumes are quantified in terms of numbers of cubes with sides of unit length, which can be called 'cubic metres', 'cubic inches', etc. For unit analysis these are better expressed in the more mathematical form of their unit symbols, for example m^3 and in^3. Nevertheless, a much-used scientific measure is the litre (L) and also such derivatives as the millilitre (mL), microlitre (μL) and decilitre (dL). The millilitre is 1 cm^3, with the abbreviation 'cc' (for cubic centimetre) perhaps more often spoken nowadays than printed. One cubic decimetre (1 dm^3) is a litre and, though cumbersome, some people advocate its use, perhaps because it relates directly to the metre. Like the decilitre, its prefix departs from the standard 'factors-of-one-thousand' sequence. Authorities express their disapproval of 'dL', 'cc', 'dm^3' and even the litre, but all continue in use.

For many people, the proper symbol for litre is a lower case L, not the capital that I am using. As noted earlier, 'l' can look like a number one with many typefaces (fonts), as well as with some handwriting.

It is easy to go wrong by several factors of 10 in converting units, but visualizing can help. A 1 L cube is readily visualized as being similar in size to a familiar 1 L flask. One can hardly then go wrong in seeing the sides of the cube as each 10 cm = 100 mm = 0.1 m long. Therefore its volume is 10 cm × 10 cm × 10 cm = 1000 cm^3, or 100 mm × 100 mm × 100 mm = 10^6 mm^3. Likewise, the volume of the 1 L cube is also $(0.1 \text{ m})^3 = 10^{-3} \text{ m}^3$ so that there must be one thousand litres in one cubic metre.

2.5 Mass

Note first that mass is not the same as weight, even though it is determined by weighing (see Chapter 14). On the basis of prefixes, one would expect the fundamental unit of mass to be the gram (g). Indeed the gram was originally defined as the mass of one cubic centimetre of water at 4 °C, the temperature at which water is densest (see Chapter 4). However, the fundamental unit of mass is now the kilogram (kg) – a fact of little practical consequence in biology. The spellings 'gramme' and 'kilogramme' are correct in ordinary English. The abbreviations 'gm' and 'kgm' are generally obsolete. The metric tonne (t), not much used in the biomedical sciences, equals 1000 kg, the approximate mass of a cubic metre of water.

2.6 Moles

> One mole (1 mol) = molar mass in grams.

The mole is the base unit for 'amount of substance'.* One mole (1 mol) of a specific homogeneous chemical substance equals its molar or molecular mass in grams. Thus 1 mol of urea is 60 g, because 60 is its molar mass. 'Molar mass' and 'molecular mass' are more correct than the commonly used phrase 'molecular weight' (MW), but sometimes 'formula mass' (or 'formula weight', FW) is preferable, as when the chemical formula does not correspond to discrete molecules, as with sodium chloride. With elements, one speaks of atomic masses or atomic weights.

When spelling out units in calculations, as, for example, in relation to concentrations (Chapter 5), one may choose to write out units of molar masses as 'g/mol' rather than just as numbers. Then, to convert 0.1 mol of urea to the mass in grams, one writes:

$$0.1 \text{ mol} \times 60 \text{ g/mol} = 6 \text{ g}.$$

Amounts of the individual ions of a salt may also be expressed in moles (mol). The molar mass of calcium chloride ($CaCl_2$) is 111. Therefore 111 g of $CaCl_2$ is 1 mol. The calcium and chloride ions are dissociated from each other in solution and those 111 g correspond also to 1 mol of calcium and 2 mol of chloride ion. The atomic mass of Ca is 40, so the 1 mol of calcium contributes 40 g to the total of 111 g.

In biochemistry and molecular biology one meets the dalton (Da), also called the 'unified atomic mass unit' (u). This corresponds to one-twelfth of the mass of carbon-12, but, putting aside details of definition, one may equate daltons numerically with molar masses. Daltons, or kilodaltons (kDa), are used particularly for proteins.

---

* 'Amount' has a special meaning here; it does not refer to mass and is not synonymous with 'quantity'.

A mole is a measure of the number of molecules or atoms, the number in 1 mol being equal to Avogadro's number. This is 6.0234×10^{23}. It is, for example, the number of oxygen molecules in 1 mol (32 g) of oxygen.

2.7 Equivalents

In the context especially of inorganic ions in solution, equivalents, or 'gram equivalents', are often just used as alternative units to moles. Therefore, more usually, milliequivalents (abbreviated variously as mequiv, m-equiv, meq, mEq and rarely mval) are alternatives to millimoles (mmol). For univalent ions (Na^+, K^+, Cl^-, HCO_3^- etc.) the two measures are the same. However, more generally, the number of milliequivalents is the number of millimoles of a given ionic species multiplied by its valency. The valency is the number of charges on each ion, used here without regard to sign ($+$ or $-$). That number is 1 in the case of the univalent cations and anions, but it is 2 for Ca^{2+}, Mg^{2+}, SO_4^{2-}, etc. Therefore, for these divalent ions, 1 mmol/L is 2 mequiv/L.

The mass of one equivalent of an ion is the atomic or molecular mass, in grams, divided by its valency. Make sure that you can reconcile this with the previous statements; it should be easy, but people can tie themselves in knots converting between moles and equivalents. Indeed, even in research papers, I have seen calcium concentrations in mequiv/L given as one-quarter of their correct values as a result of dividing millimolar concentrations by 2 instead of doubling them. One is least likely to go wrong if one thinks of milliequivalents as amounts of charge, for 1 mol of calcium (Ca^{2+}) clearly has twice the positive charge of 1 mol of sodium (Na^+).

2.1 A solution contains 40 mg of calcium per litre. The atomic mass of calcium is 40. What is the mass of 1 mequiv? What is the concentration in mmol/L and mequiv/L?

There seems to be a transatlantic difference in usage here, though I have not conducted a proper survey. In the USA milliequivalents are used for electrolytes in preference to millimoles by clinicians, physicians and pharmacists, while millimoles are preferred by such people in Europe and many other countries. Much biological, e.g. physiological, research is reported in millimoles and some first-year university chemistry courses ignore equivalents. Because of the danger of confusion (and even of non-comprehension), I would like to say that milliequivalents are best avoided for the simple purpose of recording concentrations, but that is poor advice if everyone around you, in your class or hospital perhaps, is using them.

The ionization of some substances varies with pH in its physiological range. Thus inorganic phosphate exists in plasma mainly as HPO_4^{2-} and $H_2PO_4^{-}$. One can quantify each one separately in terms of mequiv, but amounts or concentrations of total inorganic phosphate should not be given as mequiv unless the average valency is also stated.

2.8 Conversion between units

It is often necessary to convert, say, gallons to litres, yards to metres, grams to milligrams, etc. Textbooks mostly emphasize the value of working only with SI units and use these exclusively. However, practising scientists meet data that are expressed in other units. This can be tiresome and discouraging if the relevant conversion factors are not to hand, or if one is hesitant in using them.[†] Some conversion factors are tabulated in Appendix B.

Since scientists meet non-SI units, I cannot ignore them here, but SI units are used in most of the book (and such derivatives as the litre). Even then, there are conversions to be made, such as milligrams to nanograms, grams of a given chemical substance to millimoles, litres of gas to moles of gas, etc. Guidance on such conversions fits well with my theme of unit analysis.

There is a particular way of converting units that I recommend as a way of making reasoning clear and explicit, though it may seem at times unnecessary and pernickety. That is no doubt true of the following example, but it is sensible to start with a simple one.

One foot equals 0.305 metres. It requires no special tricks to see immediately that 100 feet equals $100 \times 0.305\,m = 30.5\,m$, but in order to spell out the rationale in detail one may think of the relationship between the two units in terms of the following equation:

$$\frac{0.305\,m}{ft} = 1.$$

Then,

$$100\,ft = 100\,ft \times 1 = 100\,ft \times \frac{0.305\,m}{ft} = 100 \times 0.305\,m = 30.5\,m.$$

[†] It is fortunate that we do not have to deal with solar radiation here, for, in surveying the different combinations of units used to express the latter (i.e. solar energy per unit time per unit area), I once found 17 different combinations in a mere 20 books and papers.

The point here is that if one multiplies any quantity (in this case 100 ft) by 1 it remains the same quantity. This must still be true if the '1' is replaced by an expression equal to 1, such as 0.305 m/ft. (Note, in the above working, that 'ft' is divided by 'ft' and therefore cancels out.)

In order to make other unit conversions, one may, for example, write similarly

$$1 = \frac{1000 \text{ mL}}{\text{L}}; \qquad 1 = \frac{0.454 \text{ kg}}{\text{pound}}; \qquad 1 = \frac{39.4 \text{ inch}}{\text{m}}, \text{etc.}$$

Because each of these ratios equals 1, it may also be used as its reciprocal:

$$1 = \frac{\text{L}}{1000 \text{ mL}}; \qquad 1 = \frac{\text{pound}}{0.454 \text{ kg}}; \qquad 1 = \frac{\text{m}}{39.4 \text{ inch}}, \text{etc.}$$

Furthermore, they may be used in other versions too, as convenient. The following ratios are all interchangeable, for example:

$$1 = \frac{0.305 \text{ m}}{\text{ft}} = \frac{1 \text{ ft}}{0.305 \text{ m}} = \frac{3.28 \text{ ft}}{\text{m}} = \frac{1 \text{ m}}{3.28 \text{ ft}}.$$

Such ratios may be combined, as in converting between metres and inches when the above conversion factors are not to hand:

$$1 = \frac{\text{m}}{39.4 \text{ inch}} \times \frac{12 \text{ inch}}{\text{ft}}.$$

Similarly, for converting between ounces (oz) and kilograms:

$$1 = \frac{\text{pound}}{0.4536 \text{ kg}} \times \frac{16 \text{ oz}}{\text{pound}} = \frac{35.27 \text{ oz}}{\text{kg}} = \frac{\text{oz}}{0.02835 \text{ kg}}.$$

The same method may be applied to the interconversion of moles and grams. Since the formula mass of sodium chloride is 58.5 (= 23 for Na, 35.5 for Cl), appropriate ratios are:

$$1 = \frac{58.5 \text{ g}}{\text{mol}} = \frac{58.5 \text{ mg}}{\text{mmol}} = \frac{1 \text{ mmol}}{58.5 \text{ mg}}, \text{etc.}$$

One other little point – I have chosen to write '1 mmol' as the top line of the last expression and 'mmol' without a number as the bottom line of the previous expression. Whether one writes a '1' in either position is a personal matter, for there is no rule about this.

Problems

2.1 See Section 2.7.

2.2 Physiological variables are often given in textbooks for a 'standard 70 kg man' (see Notes for Section 13.5). What is his mass in pounds (lb)? 1 lb = 0.4536 kg (Appendix B).

2.3 An inch is 25.40 mm. What is the percentage error in treating 40 inches as 1 m?

2.4 If a man is exactly 6 feet tall, what is his height in metres?

2.5 An old book gives the effective resolution of a light microscope as about 2000 Ångstrom units? What is that in micrometres (μm)?

2.6 What is the mean of 200 nm, 1 μm and 3 μm?

2.7 How many cubic micrometres (μm^3) are there in one cubic millimetre (1 mm^3)?

2.8 How many square inches are there in one square foot?

2.9 In two notoriously bad lines from an early version of his poem *The Thorn* William Wordsworth wrote, in measured metre, of a small pond:

> I've measured it from side to side:
> 'Tis three feet long, and two feet wide.

(a) Taking the pond as exactly rectangular, calculate its area in square metres. (b) If the pond had been uniformly 0.2 m deep, how many litres of water would it have held?

2.10 Express '2 mequiv Mg' in terms of millimoles.

2.11 The atomic mass of sodium is 23. The extracellular fluid of a '70 kg man' contains about 2000 mmol of sodium. What is that in grams?

2.12 The daily rate of urea excretion varies with the protein content of the diet, but we may take 400 mmol as representative. What is that in grams? The molecular mass of urea is 60.

3

Percentages

Most students know that percentages are calculated by multiplying fractions by 100 and the media assume that we all understand them. Watch out, however, for a common mistake: if you hear or read it reported that such-and-such has increased by 500 per cent, the truth may be that it has really increased *to* 500 per cent, i.e. five-fold, not six-fold as 'by 500 per cent' implies. The compulsion of the media to use percentages unnecessarily is curious. More understandable is a particularly regrettable usage in advertising. An advertisement proclaiming that a particular food is '95 per cent fat-free' does so for psychological effect, because the positive statement 'contains 5 per cent fat' puts too much emphasis on what is seen as an undesirable characteristic. Is there really a distinct fraction of the food, making up 95 per cent, that is actually fat-free?

There are other pitfalls. If the price of an item in a sale (originally £100) is reduced by 50 per cent, and then by a further 10 per cent, is the final price £50 minus 10 per cent of £100, or £50 minus 10 per cent of £50? If student A scores 60 per cent in a test and student B scores 10 per cent more, does student B score 70 per cent or 66 per cent?

Percentages are generally unit-free, or dimensionless, numbers, since the units involved in their measurement should cancel out in calculation. Thus if a 90-kilogram individual contains 54 kilograms of water, the percentage of water is

$$\left(100 \times \frac{54 \text{ kg}}{90 \text{ kg}}\right) = 60\%.$$

Biomedical Calculations   Richard F. Burton
© 2008 John Wiley & Sons, Ltd

The answer would be the same if the masses were expressed in grams, ounces or pennyweight. Note, however, that percentages can also be given in terms of volumes and the resulting numbers may not be the same as those calculated from masses. Thus, if a mixture of ethanol and water contains 50 per cent ethanol in terms of mass, the percentage in terms of volume (e.g., millilitres of ethanol in millilitres of mixture) is 58 per cent, because alcohol has a lower density than water. The percentage may also be expressed in terms of grams per 100 mL, in which case it is 46 per cent. The three possibilities are commonly distinguished as 'w/w', 'v/v' and 'w/v', standing respectively for 'weight/weight', 'volume/volume' and 'weight/volume'. Especially with w/v, it is advisable to write out the units, e.g. g/100 mL, when calculating or performing unit analysis.

3.1 When percentages mislead: human body fat and fat in milk

Coming back to science, let us look now at another situation where percentages can be misleading. The percentage of fat in the human body obviously varies. Consider a fairly lean person containing, say, 15 per cent fat and a very obese one with a fat content of 60 per cent. Can one say that the second contains four times as much fat as the first (because 60/15 = 4) or is 'four times as fat'? The question is badly framed, because it is unclear whether the two total body masses are the same, or the body masses minus the fat are the same, or neither. However, the answer is 'no' anyway. It is worth exploring the situation in detail, because a significant and general point about the use of percentages will emerge.

The body may be thought of as having two components, a fat-free mass, N, and mass of fat, F, making the total mass $(F + N)$. The percentage of fat is $100F/(F + N)$. N and F should have the same units and these cancel to give the percentage as a unit-free number. If a hypothetical body of constant fat-free mass lays down fat, the percentage does not increase exactly in proportion to the mass of fat F, but increases instead proportionately with $100F/(F + N)$. This is not a straight-line relationship.

3.1   Consider two individuals, with each now specified as having a fat-free mass of 60 kg. One contains 15 kg of fat and the other contains 90 kg (i.e. six times as much). What is the percentage of fat in each individual? Is one percentage also six times the other?

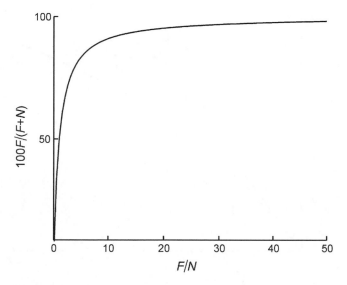

Figure 3.1 If a body or other object consists of two components, of masses F and N, the percentage of F is 100 $F/(F + N)$. The graph shows the relationship between this and the ratio F/N. If N is constant, the graph may be read as showing the relationship between the absolute and percentage contents of component F

Human fat contents are generally stated as percentages. That may often be satisfactory, but for some statistical and modelling purposes in research it may be better to use the ratio F/N. The relationship between percentage, i.e. $100F/(F + N)$, and F/N is shown in Figure 3.1. The curve applies to anything that can be described in terms of two components and it is for this reason that the caption does not actually state what F and N represent.

The curve in Figure 3.1 is distinctly nonlinear, but at low values of F/N it is much straighter. This is shown in Figure 3.2, in which the lower left part of the curve of Figure 3.1 is magnified.

3.2   If you have many values for fat contents expressed as F/N and need to convert them to percentages, it might help to have a general formula for doing this. Derive one, starting from the equation:

$$\text{percentage} = 100F/(F + N).$$

The following two problems illustrate the same nonlinearity, but refer to the fat content of milk rather than of bodies. For simplicity, regard the percentages as

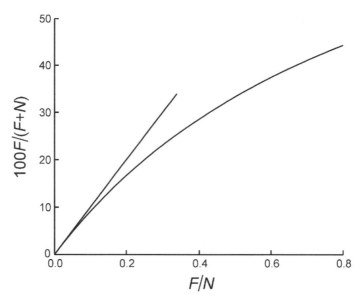

Figure 3.2 This shows the lower part of the curve of Figure 3.1 magnified. At low values of F/N, the curve is more nearly linear

being calculated for masses of milk and fat (i.e. 'w/w') rather than for volumes ('v/v') or a combination of both ('w/v'). Containers of milk for domestic use are usually labelled in terms of w/v. The fat content of human milk averages about 4.5 per cent.

3.3 Compare cow's milk containing 4 per cent fat and rhinoceros milk containing 0.3 per cent fat. The ratio of these two percentages is 13.3. The value of F/N for the cow milk is 0.0417, calculated as $4/(100-4)$. What is the equivalent value for the rhinoceros milk? What is the ratio of the two values of F/N? Compare the answer with the ratio of percentages.

3.4 In the same way compare milk from a grey seal containing 53 per cent fat with cow's milk containing 4 per cent fat. Now the ratio of the percentages is 13.25, almost the same as for the cow and the rhinoceros. The value of F/N for the cow milk is again 0.0417. What is the equivalent value for the seal milk? What is the ratio of the two values of F/N? Again compare the answer with the ratio of percentages.

> **3.5**   In the above two problems, which stated numbers are pointlessly over-precise?

It should not be concluded that the use of percentages is generally inappropriate. Given milk containing 4 per cent fat, one might wish to know the absolute amount of fat contained in 500 g. The calculation is easy: $4/100 \times 500$ g $= 20$ g. Aside from the ease of the arithmetic, it is simpler to talk about percentages than 'ratios of fat to non-fat'.

> **3.6**   Is it as easy to calculate the amount of fat in a given quantity of milk from the ratio, F/N? If this ratio is 0.05, how much fat is there in 500 g of milk?

3.2 Heat loss from the body: further questionable percentages

Heat loss from the body occurs by radiation, convection (warming of air as it passes over the general body surface and through the airways) and evaporation of water from the same surfaces. The relative amounts of these three kinds of heat loss vary with circumstances and the total loss increases with metabolic heat production during exercise. Blood is diverted to the skin. Sweating increases. More air is expired per minute. Activity increases the flow of air over the body surface. Table 3.1 shows some representative values for total heat loss and for the percentages of those totals that are lost by radiation, convection and evaporation. Look at the table and describe to yourself in words rather than numbers what is happening. For the moment, kilocalories need only be understood vaguely as units of heat energy.

Table 3.1 Heat loss from a person at rest and during vigorous exercise, after Du Bois (1938) and Burton (1998)

	Rest	Exercise
Total heat loss, kilocalories per hour	70	700
Radiation	65%	15%
Convection	15%	15%
Evaporation	20%	70%

> **3.7** Calculate the actual rates of heat loss at rest and in exercise in kilocalories per hour. How well do the results agree with your verbal description?

Problems

3.1 See Section 3.1.

3.2 See Section 3.1.

3.3 See Section 3.1.

3.4 See Section 3.1.

3.5 See Section 3.1.

3.6 See Section 3.1.

3.7 See Section 3.2.

3.8 A man weighing 70 kg has a brain of 1300 g. Express the brain size as a percentage of body mass.

3.9 An individual produces 1100 mL of urine in a day containing 65 g of solid material as determined by drying. What is the percentage of solids (w/v)?

3.10 A particular adult of 70 kg contains 62 per cent water and 15 per cent fat. What is the mass of all the other material?

3.11 Consider a hypothetical 70 kg person, initially containing 62 per cent water and 15 per cent fat as above. (a) What is the percentage water content of the non-fat component? (b) Suppose that 8 kg of fat are laid down with no change in the other masses. What is the final percentage water content?

3.12 In a sample of blood 63 per cent of the leucocytes are granulocytes and 4.5 per cent are monocytes; 4 per cent of the granulocytes are eosinophils. There are 7400 leucocytes per microlitre. What percentage of the leucocytes consists of eosinophils? How many eosinophils are there per microlitre?

4

Composite units I – density

The density of an object or substance is its mass per unit volume, calculated as mass divided by volume. It has such composite units as $g\,mL^{-1}$, $kg\,L^{-1}$ and $kg\,m^{-3}$. Densities are needed for the next topic, namely that of concentrations, but we also look here at two definitions of body density, one relating to buoyancy and the other to the estimation of body fat.

4.1 Of the above units ($kg\,m^{-3}$, $g\,mL^{-1}$ and $kg\,L^{-1}$), which two are numerically identical?

$$\text{Density} = \frac{\text{mass}}{\text{volume}}$$

The density of pure water is exactly $1\ g\ mL^{-1}$ at 4 °C. It is maximal at that temperature, but it is always within 0.8 per cent of that value over the temperature range 0–40 °C. Therefore $1\ g\ mL^{-1}$ is a satisfactory value to remember for most purposes. (Ice, which floats on water, has a density of $0.915\ g\ mL^{-1}$.)

Densities may be expressed in a variety of combinations of mass and volume units, but usually these are chosen in a way that makes the density of water numerically close to unity, e.g. $g\ mL^{-1}$ or $kg\ L^{-1}$. However, there is nothing wrong with, say, $g\ L^{-1}$. The choice may depend on the mass and volume units that otherwise appear in a calculation – with tonne m^{-3} possibly preferred for building materials, for example.

Biomedical Calculations   Richard F. Burton
© 2008 John Wiley & Sons, Ltd

4.1 Specific gravity

Densities are sometimes expressed as 'specific gravities'. The specific gravity of a substance is its density divided by the density of water, the water usually being at $4\,°C$, the temperature at which its density is maximum. Specific gravities have no units. To specify the temperatures of the substance in question and of the water to which the substance is being compared, a symbol of the form 'd_4^{20}' may be used. In this case the substance is at $20\,°C$ and the water is at $4\,°C$. In the case of 'd_4^{20}' the density is numerically identical to the specific gravity, because the density of water at $4\,°C$ is exactly 1 g mL^{-1}.

　　Specific gravities are sometimes used to gauge the concentrations of solutions. Thus the specific gravity and density of wine (measured with a hydrometer) is affected by the content of sugar and alcohol, with sugar increasing the specific gravity and alcohol lowering it. There is also a simple method of judging the specific gravity of blood, and thus the concentration of haemoglobin in it, by seeing whether drops of blood sink or float in copper sulphate solutions of different concentration. This is widely used in blood donor screening. The specific gravity of plasma, d_{20}^{20}, is typically about 1.025–1.029. Its variability depends mainly on the protein content and can be used as a measure of this (see below).

4.2 Specific volume

In some contexts it is useful to use the reciprocal of density, that is to say the volume for a given mass (e.g., mL g^{-1}). This is known as the 'specific volume'. The plasma proteins are more dense than water and protein-free plasma and their mean specific volume is therefore less than 1 mL g^{-1}. It is in fact about 0.73 mL g^{-1}. This is referred to as their partial specific volume, because 0.73 mL is the increase in the total volume of a solution when one gram of protein is added. From it one may estimate the contribution of proteins to the total volume of a plasma sample (see Notes).

　　Let us use this value for the partial specific volume of plasma proteins to calculate the relationship between the protein concentration of plasma and the plasma density. We also need to know the density of protein-free plasma – and that is 1.007 g mL^{-1}. Let us take the example of 100 mL of plasma containing exactly 7 g of protein (i.e. 7 g/100 mL, 0.07 g mL^{-1}). The contribution of the 7 g of protein to the total volume of 100 mL is 7 g $\times 0.73$ mL $g^{-1} = 5.11$ mL. The volume of the rest (equivalent to protein-free plasma) is therefore $(100 - 5.11)$ mL $= 94.9$ mL. The mass of this is 94.9 mL $\times 1.007$ g $mL^{-1} = 95.6$ g. The total mass of the 100 mL of plasma is therefore $(95.6 + 7)$ g $= 102.6$ g. The density is 102.6 g/100 mL $= 1.026$ g mL^{-1}.

For repeated routine calculations of protein concentration from density, a formula would be quicker. To obtain this, one may replace the '7' in the above calculation with a symbol, say P, with this being the number of grams of protein in 100 mL.

4.2 Derive an equation relating P to the density, D, of a plasma sample.

4.3 Two definitions of body density

The density of body fat is markedly less than that of the rest of the body. Therefore the density of the body decreases with increasing fat content. This is of interest in two important contexts. Firstly, by affecting buoyancy, it matters both to swimmers and to those unfortunates who are 'not waving but drowning'. Secondly, exact measurements of body density can be used in the estimation of body fat. Here we look at both these things. The calculations run in both directions – from body composition to density and from density to body composition.

Body density depends not only on fat content, but on the proportions of bone mineral, soft tissue and the air or gas in the lungs, airways and gut. Here, however, we treat the body as made up just of three components, namely fat, other tissues (including bone mineral) and air. The non-fat tissues constitute what is called the 'fat-free mass'. The term 'lean body mass' is sometimes used to mean the same thing, but it also sometimes excludes the bone mineral.

The density of fat is 0.90 g/cm^3 (or g mL^{-1}, kg/L, etc). The density of the fat-free mass (excluding air) varies a little with the proportion of bone mineral present, but is usually taken as 1.10 g/cm^3 or 1.10 kg/L. The density of air is low enough to be taken as zero in this context, but its volume is neither negligible nor constant during breathing.

Consider an individual consisting of a fat-free component of 60 kg, 22 kg of fat and, at this particular instant, 3 L of air. What is the body density? We need to know the total mass and the total volume. The total mass is 60 kg plus 22 kg, namely 82 kg. As for the volume, we know that of the air, but need to calculate those of the two other components from their masses and densities. Let us start with the fat-free component. We can either stop to think or we can write down the two quantities with their units so that it becomes obvious what we might do with them:

$$60 \, \text{kg} \qquad 1.10 \, \text{kg/L}$$

The answer needs to be in litres and should not include 'kg'. Thus, to achieve appropriate units, we must divide kg by kg/L:

$$60 \, \text{kg} \times \frac{\text{L}}{1.10 \, \text{kg}} = 54.5 \, \text{L}.$$

Calculated in the same way, the volume of fat is:

$$22 \, \text{kg} \times \frac{\text{L}}{0.90 \, \text{kg}} = 24.4 \, \text{L}.$$

The total volume is therefore $(54.5 + 24.4 + 3) \, \text{L} = 81.9 \, \text{L}$. The overall body density, is therefore,

$$\frac{82 \, \text{kg}}{81.9 \, \text{L}} = 1.00 \, \text{kg/L, or } 1.00 \, \text{g/cm}^3.$$

Obviously the numbers have been chosen to give this individual the density of fresh water. The fat content in this case is $100 \times 22 \, \text{kg}/(60 + 22) \, \text{kg} = 27\%$, so that that is the fat content conferring neutral buoyancy in fresh water (but changing slightly during breathing). (Sea water is more dense, so this individual would definitely float in the sea.) Obesity is sometimes defined as a fat content exceeding 20 per cent in men and 25 per cent in women; the 27 per cent thus represents a small degree of obesity.

Now we come to the estimation of body fat from density. For this purpose the density is taken to be that of the body less the air in the lungs and airways. The body mass is virtually the same whether or not this air is included, but the volume now being considered does not include that volume of air. We are thus dealing with a different density now. It is somewhat higher than the overall density determining the buoyancy of bathers. Note this distinction because it sometimes causes confusion.

The density, which has to be accurately known for estimating body fat, can be measured by weighing the body both in air and submerged in water. An adjustment is then made for the air in the lungs and airways so that the body is treated as air-free. (We will not concern ourselves with the technicalities of doing that.) Calculating from first principles is slightly more awkward than the previous calculation and is dealt with in Chapter 6. However, there are simple formulae for routine calculations and we look next at one of these.

4.4 Thinking about a formula

There is a commonly used formula (the Siri equation) for calculating percentage body fat from body density (see Notes). In this context 'body density' refers to the body with air excluded (see above). Here is the formula as quoted in one particular publication:

$$\text{percentage body fat} = (495\,\text{g/cm}^3)/(\text{density}) - 450. \qquad (4.1)$$

This formula is only useful for readers who are in the special position of having the means of measuring body density, but we may try to make sense of it as an exercise in quantitative thinking.

Looking first at units, we note that 'percentage body fat' and '450' have none. Is that true of the remaining term? The density units are not stated, but one might guess that they would be 'g/cm^3'; in fact that is what they must be if they are to cancel and leave '(495 g/cm^3)/density' unit-free like the other terms.

The formula can be derived from first principles, assuming appropriate densities for the two components of the body. However, let us do instead two quick tests of the equation. First consider a hypothetical individual who is completely fat-free. That person's density would be 1.10 g/cm^3 (see above). Inserting that value into Eq. (4.1) gives a percentage fat content of $(457/1.1 - 450) = 0\%$. That is as it should be, so this test of the equation is satisfactory. Next consider someone with a density of 0.90 g/cm^3, which is the same as pure fat (an impossible person of prodigious obesity). This time the percentage fat works out at $(495/0.90 - 450) = 100\%$. Again this is appropriate. Even more easily, we can calculate the percentage fat content of an individual with a density of 1.00 g/cm^3; the answer is $495 - 450 = 45\%$. This is higher than the 27 per cent calculated above for a person of that density, but in that case the individual contained 3 L of air. If we had approached the formula with any scepticism, we may now feel reassured.

Problems

4.1 See the beginning of this chapter.

4.2 See Section 4.2.

4.3   Which of the following are the same density as 1.1 g/mL?

$$1.1\,\text{g/cm}^3, \; 1.1\,\text{kg}\,\text{L}^{-1}, \; 1100\,\text{g/L}, \; 1.1 \times 10^6 \,\text{g}\,\text{m}^{-3}.$$

4.4   Using Eq. (4.1), calculate the percentage fat content of an individual with density (appropriately measured) of 1.05 g/cm³.

4.5   A 2 per cent (w/w) solution of sodium chloride has a specific gravity (d_4^{20}) of 1.0125. What is the concentration in g L^{-1}?

5
Composite units II – concentration

$$\boxed{\text{Concentration} = \frac{\text{Quantity of solute}}{\text{Quantity of water or solution}}}$$

Making up and diluting solutions are common activities – in relation, for example, to physiological and microbiological media, drug solutions and biochemical experiments. In addition, concentrations both in body fluids and in artificial solutions may need to be re-expressed (e.g. with grams converted to millimoles, litres of solution converted to kilograms of water, etc.). Attention to units, i.e. unit analysis, is therefore again important. Here I deal with masses, millimoles and volumes, deferring detailed treatment of milliequivalents, milliosmoles and activities to Chapter 9.

Concentrations of aqueous solutions are often given in terms of mass of solute (i.e. dissolved substance) per unit volume of solution. The units might then be, for example, $g\ L^{-1}$ or $\mu g\ mL^{-1}$. As a general formula:

$$\text{Concentration} = \frac{\text{Mass of solute}}{\text{Volume of solution}}. \qquad (5.1)$$

Sometimes concentrations are expressed in terms of the mass or volume of water, rather than the volume of solution. They are also sometimes given as percentages, without mention of masses or volumes. Thus one may meet a

Biomedical Calculations   Richard F. Burton
© 2008 John Wiley & Sons, Ltd

'5 per cent solution' of, say, sucrose. On its own, the phrase '5 per cent solution' is ambiguous, but it may be accompanied by the abbreviation 'w/w' or 'w/v'. As noted in Chapter 3, these indicate respectively 'weight of solute/weight of solution' (w/w) and 'weight of solute/volume of solution'. (That 'weight' is better rendered as 'mass' has already been indicated.) Spelt out in appropriate units, the 5 per cent sucrose would be either '5 g sucrose/100 g solution' (w/w) or '5 g sucrose/ 100 mL solution' (w/v). When two liquids are mixed (e.g. alcohol and water), the abbreviation 'v/v' may be used when the volume of one component is being expressed as a percentage of the total volume.

The solute component in a solution may be specified in terms of moles, millimoles, micromoles, etc., rather than of mass. A solution containing 100 millimoles per litre of solution (100 mmol L^{-1}, 100 mmol/L, 0.1 mol L^{-1}, 0.1 mol/L) may be described as 100 millimolar or 0.1 molar. Alternative abbreviations for the latter are 100 mM and 0.1 M. Concentrations may again be specified in terms of kilograms of water, e.g. as 100 mmol/kg water, etc. Then the words 'molar' and 'millimolar' are replaced by 'molal' and 'millimolal'. One may speak of the 'molarity' or 'molality' of a solution. Writing 'M' or 'mM' is convenient, but 'mol/L' and 'mmol/L', or 'mol L^{-1}' and 'mmol L^{-1}', are generally more appropriate for unit analysis.

> 'mM' means 'mmol/L' or 'mmol L^{-1}'.
> 'mM' is less appropriate for unit analysis.

Dissolved gases may be quantified in these terms, but concentrations of oxygen and carbon dioxide in biological fluids are most commonly expressed as volumes of gas per unit volume of fluid. The gas volumes are what the gases would occupy if released from solution and measured at standard temperature and pressure (see Chapter 17). Concentrations of oxygen or carbon dioxide in blood, as usually stated, include components chemically combined as well as in simple solution.

Units stated baldly as mass/volume, e.g. g/L, could also correspond to density (Chapter 4). Confusion is unlikely in most contexts, but for unit analysis it is often helpful to specify what the mass and volume refer to – e.g. g salt/L solution. In the following paragraphs it is particularly important to specify what the masses and volumes refer to, whether solutes, solutions or water.

5.1 Concentrations: kilograms of water vs litres of solution

With simple solutions of salts at ordinary biological concentrations there is typically little difference numerically between a concentration expressed as a quantity of solute per litre of solution (e.g. molarity) and the same concentration expressed as a quantity of solute per kilogram of water (e.g. molality). With blood plasma the differences are more significant, especially in the context of the smallness of natural variations in, say, sodium. This is because of the presence of plasma proteins and lipids which contribute significantly to the total volume, but not to the mass of water. Thus a plasma sodium concentration of 141 mmol/L might correspond to 150 mmol/kg water. The difference with cytoplasm is much greater, because of its higher content of space-occupying organic materials.

5.2 Simple protein-free salt solutions

As already stated, there is little difference between the molarities and molalities of simple, dilute salt solutions. The calculations below are to illustrate this point. Such calculations are rarely required, but they do illustrate the usefulness of spelling out what units refer to (i.e. as an exercise in unit analysis). They also lead to a reassuring conclusion.

In exploring this through calculation we cannot assume that the volume of a solution equals the sum of the separate volumes of the components before mixing. Thus there is significant volume shrinkage when water and ethanol are mixed. To consider a specific and more relevant case, let us suppose that exactly 10 g of salt (NaCl) is dissolved in exactly 1 kg of water at 20 °C, giving a concentration of 0.0100 kg NaCl/kg water ($=$171 mmol/kg water), and ask what the concentration is when expressed as kg NaCl/L solution.

The mass of solution containing 1.0000 kg of water is:

$$1.0000 \text{ kg water} + 0.0100 \text{ kg salt} = 1.0100 \text{ kg.}$$

Therefore,

$$\text{kg solution/kg water} = 1.0100.$$

According to published tables, the density at 20 °C is 1.0053 kg/L solution. Thus,

$$\text{kg solution/l solution} = 1.0053.$$

Dividing the second ratio by the first, we get:

$$\frac{\text{kg solution/L solution}}{\text{kg solution/kg water}} = \frac{1.0053}{1.0100} = 0.9953.$$

Thus

$$\text{kg water/L solution} = 0.9953.$$

(As a check, ask yourself whether, for any solution of NaCl, you would expect a value less than 1.) The last expression can be rearranged as

$$\frac{\text{kg water}}{0.9953 \text{ L solution}} = 1.$$

Therefore, at 20 °C, the concentration of the solution (0.0100 kg NaCl/kg water) can be re-expressed as

$$\frac{0.0100 \text{ kg NaCl}}{\text{kg water}} \times \frac{\text{kg water}}{0.9953 \text{ L solution}} = 0.01005 \text{ kg NaCl/L solution.}$$

In this case there is virtually no numerical difference between the two measures. We may have strained to bring forth a gnat, but from now on we can be relaxed about the matter.

5.3 Millimolar and millimolal concentrations in blood plasma

For readers who are concerned about the difference between millimolar and millimolal concentrations in blood plasma, here is how to make the conversion. We start by defining a conversion factor, f, such that

$$f = \frac{\text{mmol/L solution}}{\text{mmol/kg water}} = \text{kg water/L solution.}$$

Let the protein concentration be c g/L and let f have a value of f_0 when c is zero. For plasma that is free of proteins, f_0 is about 0.99 kg water/L solution. We also need to consider the volume, V, occupied by a gram of protein (called its 'partial specific volume' – see Chapter 4). For many proteins V is 0.70–0.75 mL/g and for plasma proteins in particular it is about 0.73 mL/g. The following formula shows how to calculate f:

$$f = f_0(1 - Vc/1000) \qquad (5.2)$$

or, specifically for blood plasma,

$$f = 0.99(1 - 0.73c/1000).$$

The concentration of plasma proteins, c, is typically about 60–84 g/L. For 70 g/L, the value of f is 0.94. The normal concentration range for sodium is about 137–145 mmol/L. A mid-range value of 141 mmol/L, with $f = 0.94$, corresponds to 141/0.94 = 150 mmol/kg water. This convenient round number quantity is almost exactly matched in protein-free solutions (artificial salines and glomerular filtrate) by a millimolar concentration of 150 mmol/L (see above).

5.4 Some quite different uses for Eq. (5.1)

One reason for stating the relationship of Eq. (5.1) as a formula is that it can be put to other uses. Rearranged, it becomes:

$$\text{Mass of solute} = \text{concentration} \times \text{volume of solution} \qquad (5.3)$$

and

$$\text{Volume of solution} = \frac{\text{mass of solute}}{\text{concentration}}. \qquad (5.4)$$

For practice you may wish to try out unit analysis on both new versions. Remember that the concentrations here are 'masses per unit volume'.

Equation (5.3) can be used in calculating the dose of a drug administered in a solution of known volume and concentration. Equation (5.4) may be used in a completely different context – for estimating volumes of fluid that are not accessible for measurement in other ways. Thus the volume of extracellular fluid in the body may be estimated by injecting into it a known amount of a substance that does not enter the cells, allowing it to be thoroughly mixed within the extracellular fluid, and then determining its final concentration. A suitable substance, much used in the past, is the plant polysaccharide inulin. The principle is simple, but the procedure is complicated by the need to allow for some excretion in urine during the period of mixing (see Chapter 19).

Problems

See Chapters 7 and 8 for calculations relating to the preparation of solutions and on drug dosage.

5.1 With regard to potassium in plasma, which is greater, the concentration in mmol/L or the concentration in mmol/kg water?

5.2 Why are concentrations in cells and tissues usually given as molalities rather than as molarities?

5.3 Drug Y is available as a solution of concentration of 500 mg L^{-1}. What volume must be administered to a patient if the required dose is 4 mg? How many ways can '500 mg L^{-1}' and '4 mg' be sensibly combined on their own in a calculation?

5.4 In regard to calcium, which of the following are not the same as 2 mmol/L? 2 mM, 0.2 mmol dL^{-1}, 2 mequiv/L, 4 mequiv L^{-1}, 1 mEq/L, 0.2 mmol/100 mL.

5.5 The water content of a dead mouse may be found by weighing it before and after drying, but this method is unsuitable for living animals or people. Instead, one may inject some substance that is known to mix freely with all the body water, and then find its final concentration. Suitable substances are antipyrine and heavy or tritiated water: 500 mg of antipyrine are injected into a human subject and after equilibration the concentration is found to be 8.6 mg/L water. How much water is there in the body?

5.6 Sometimes a line of reasoning can be applied in a completely different context, with resulting economy of thought. Consider a well-defined and stable population of geese in a certain area. The size of the population is unknown. Thirty geese are captured, ringed and released to mingle with the rest. Subsequently, 100 are captured and 10 are found to have been ringed on the previous occasion. What is the likely size of the population? To which of the above equations does this problem most closely relate?

6

Aspects of problem solving

6.1 Letting unit analysis solve the problem

In Chapter 1, I showed how the answer to a particular problem might be found
without either understanding it or even knowing what it is. The only clues were
two items of data ('2 gram NaCl' and '0.1 litre water'). Admittedly the problem
was a very easy one, but it did illustrate an important point. Often, but not always
of course, if you pay attention to the units, they propel you towards the answer.
This is a supplement, or even an alternative, to the usual processes of reasoning.

> With attention to the units some problems solve themselves.

Let us now see the method in action with the following more complicated problem.

> The molecular mass of haemoglobin (Hb) is 64 650. If a sample of
> blood contains 150 g of haemoglobin per litre and the erythrocytes
> make up 44 per cent of the blood by volume, what is the concentration
> of haemoglobin in mmol per litre of erythrocytes?

There are three quantitative statements there and the first step is to express them
as ratios, each equal to one. Here they are:

A $\quad \dfrac{64\ 650 \text{ g Hb}}{1 \text{ mol Hb}} = 1.$

Biomedical Calculations   Richard F. Burton
© 2008 John Wiley & Sons, Ltd

B $$\frac{150 \text{ g Hb}}{1 \text{ L blood}} = 1.$$

C $$\frac{44 \text{ L erythrocytes}}{100 \text{ L blood}} \quad \text{or} \quad \frac{0.44 \text{ L erythrocytes}}{1 \text{ L blood}} = 1.$$

Notice now that haemoglobin is represented in the problem in terms both of a mole and as millimoles (in 'mmol/L'). A conversion factor is therefore needed, namely either

D $$\frac{1000 \text{ mmol Hb}}{1 \text{ mol Hb}} = 1 \text{ or}$$

E $$\frac{1 \text{ mol Hb}}{1000 \text{ mmol Hb}} = 1.$$

One could write these two out without specifying 'Hb', but being specific can slightly help in the later stages of the calculation.

The next step is to combine these ratios in a single expression, multiplying or dividing as appropriate, and perhaps choosing to write the ratios down as their reciprocals (i.e. inverting them). What is required is that most of the units should cancel out to leave only those required in the answer. In this case the units of the answer need to be

$$\frac{\text{mmol Hb}}{\text{L erythrocytes}}.$$

So here goes. To get 'mmol Hb' on the top line, we can start with the first of the above ratios (A), but inverting it. However, either D or E is needed too, in order to convert 'mol Hb' to 'mmol Hb'. Thus:

$$\frac{1 \text{ mol Hb}}{64\ 650 \text{ g Hb}} \times \frac{1000 \text{ mmol Hb}}{1 \text{ mol Hb}}$$

or

$$\frac{1 \text{ mol Hb}}{64\ 650 \text{ g Hb}} \div \frac{1 \text{ mol Hb}}{1000 \text{ mmol Hb}}.$$

I find it slightly more straightforward to multiply these ratios, rather than divide, so let us choose the first version. We could simplify the expression at this stage, but let us instead write out the whole calculation in one line, like this:

$$\frac{1 \text{ mol Hb}}{64\ 650 \text{ g Hb}} \times \frac{1000 \text{ mmol Hb}}{1 \text{ mol Hb}} \times \frac{100 \text{ L blood}}{44 \text{ L erythrocytes}} \times \frac{150 \text{ g Hb}}{1 \text{ L blood}}$$

Before concerning ourselves with the numbers, we now check to see whether the units cancel appropriately to leave us with 'mmol Hb/L erythrocytes'. They do. If they had not done so, then inverting one or more of the ratios might have been necessary. If that did not work, then something else would have gone wrong, with possibly another ratio being needed.

However, all is well. The next step is to do the arithmetic on the numbers:

$$\frac{1}{64\ 650} \times 1000 \times \frac{100}{44} \times 150 = 5.3$$

So the answer is 5.3 mmol Hb/L erythrocytes.

Note that the order in which the ratios are written down is unimportant, so long as one can see that 'mmol Hb' will eventually remain on the top line after the cancelling. There is no need to think through, and write down, the calculation as separate acts of multiplication or division, though one may prefer to do that. All that is required is to write down the relevant ratios (all equal to 1), write down the units of the answer in ratio form and then write down the other ratios as a string of multiplications, inverted as necessary, so that cancellations leave just the required units.

With real-life problems, as opposed to most book exercises, there may be more available information than is required, or not enough. This may become apparent at the stage when the ratios have been assembled for the final calculation, for then there would be units in one or more of the ratios that are neither needed in the answer, nor capable of being cancelled.

We return to this method of calculating later, in the contexts of making up solutions (Chapter 7) and calculating drug doses (Chapter 8). It is not appropriate to all problems, of course. Simple examples include those requiring addition or subtraction. However, it may still be helpful to express some of the relevant information in ratio form (including conversion factors). We look next at problems where it helps to represent the answer by a symbol and then to apply simple algebra.

6.2 'Let *x* be the unknown'

The calculations so far have mostly been straightforward. We look now at problems where some rearrangement of equations is involved and where, for that purpose, it helps (if only for speed and neatness) to represent the answer by a symbol. Traditionally '*x*' or '*X*' is used (preferably not written as × in case it is mistaken for a multiplication sign – besides, italics are usual for such symbols when printed). However, it is sometimes better to use something more suggestive of what is represented – perhaps an abbreviation or an initial letter. The chosen symbol may represent both the numerical component and the unit component of the unknown quantity that is to be determined, or sometimes it may just stand for the unknown number, without its units. Here are five examples.

Representing the required answer by means of a symbol can be helpful if some rearrangement of terms in an equation is needed.

---

1. Suppose that you need to prepare a 2 mM solution of calcium chloride in a 250 mL volumetric flask. You have a graduated pipette and a 100 mL flask containing calcium chloride at a concentration of 100 mmol/L. How much of the latter solution is needed for dilution to a final volume of 250 mL? One way to work out the answer in an orderly fashion on paper is to call that volume *x* mL and write:

$$2 \text{ mmol/L} = 100 \text{ mmol/L} \times \frac{x \text{ mL}}{250 \text{ mL}}.$$

This now needs to be re-arranged so that the *x* is on the left side, on its own. One way of achieving this, writing out all steps to avoid mistakes, is as follows. First swap over the two sides.

$$100 \text{ mmol/L} \times \frac{x \text{ mL}}{250 \text{ mL}} = 2 \text{mmol/L}.$$

Then divide both sides by 100 mol/L and multiply both sides by 250 mL.

$$\frac{100 \text{ mmol/L}}{100 \text{ mmol/L}} \times \frac{x \text{ mL}}{250 \text{ mL}} \times 250 \text{ mL} = \frac{2 \text{ mmol/L}}{100 \text{ mmol/L}} \times 250 \text{ mL}$$

After the obvious cancellations, we are left with:

$$x \text{ mL} = 2/100 \times 250 \text{ mL} = 5 \text{ mL}.$$

2. A related problem is to work out how to prepare a 5 mM solution of potassium chloride (molecular mass 74.6) in a 250 mL volumetric flask, starting with the dry salt. Here we may let x stand for the required mass of KCl in grams and calculate as follows:

$$\frac{x \text{ g}}{250 \text{ mL}} = \frac{5 \text{ mmol}}{\text{L}} \times \frac{1 \text{ L}}{1000 \text{ mL}} \times \frac{1 \text{ mol}}{1000 \text{ mmol}} \times \frac{74.6 \text{ g}}{\text{mol}} = \frac{0.373 \text{ g}}{1000 \text{ mL}}.$$

We find the answer, x g, by multiplying both sides by 250 mL:

$$x \text{ g} = \frac{0.373 \text{ g}}{1000 \text{ mL}} \times 250 \text{ mL} = 0.09325 \text{ g}.$$

This calculation may seem long-winded, especially to someone who makes up solutions frequently, but, even if much of the reasoning takes place in the head, the fundamental elements of the calculation may be identical. Actually there is no need here to involve an 'x' at all and you may care to find for yourself an alternative, x-free, version of the calculation (see Notes). No harm is done if one postulates an 'x' unnecessarily; one should do what comes most easily.

3. This example takes us back to the estimation of body fat from density (Chapter 4). In this case we take body density as

$$\frac{\text{Body mass}}{\text{Body volume, airway air excluded}}$$

$$= \frac{\text{Body mass}}{\text{Fat volume} + \text{volume of remainder, air excluded}}.$$

Let us say that the total body mass is 75 kg and that its density has been measured as 1.03 kg/L. As earlier, the fat density may be taken as 0.90 kg/L and the density of the fat-free mass may be taken as 1.10 kg/L. How many kilograms of fat does this body contain? Let us represent the answer (i.e. the number without its unit) by the symbol F. Then the remainder, the fat-free mass, is $(75 - F)$ kg.

We obtain the respective volumes by dividing by the densities:

$$\text{Volume of fat} = \frac{F \text{ kg}}{0.90 \text{ kg/L}} = 1.111 F \text{ L.}$$

$$\text{Volume of remainder} = \frac{(75 - F) \text{ kg}}{1.10 \text{ kg/L}} = (68.2 - 0.909 F) \text{ L.}$$

Thus the total volume is $1.111 F$ L $+ (68.2 - 0.909 F)$ L.
Now we can equate two expressions for the body density:

$$\text{Body density} = 1.03 \text{ kg/L} = \frac{75 \text{ kg}}{(1.111F + 68.2 - 0.909F) \text{ L}}.$$

Therefore,

$$1.03 \text{ kg/L} = \frac{75}{(0.202F + 68.2)} \text{ kg/L.}$$

The units are the same on both sides of the equation, so can be cancelled out and dropped. If both sides are now multiplied by $(0.202\,F + 68.2)$,

$$1.03 \times (0.202F + 68.2) = \frac{75}{(0.202F + 68.2)} \times (0.202F + 68.2)$$

Therefore

$$(1.03 \times 0.202) F + (1.03 \times 68.2) = 0.208 F + 70.2 = 75.$$

Rearrangement then gives:

$$0.208 F = 75 - 70.2 = 4.8,$$

so that

$$F = 4.8/0.208 = 23.1.$$

Therefore the mass of fat is 23.1 kg. This corresponds to a percentage fat content of $(23.1 \text{ kg})/(75 \text{ kg}) \times 100 = 31\%$.

6.1   Check this percentage using the Siri equation given in Chapter 4:

$$\text{percentage body fat} = (495 \text{ g/cm}^3)/(\text{density, g/cm}^3) - 450. \qquad (4.1)$$

(This equation is derived by similar reasoning, but using algebra instead of specific numbers.)

---

4. Suppose that you have a 45 g tube of 2.5 per cent hydrocortisone cream and a 30 g tube of 1 per cent hydrocortisone cream and that you need to prepare from these a 30 g tube of 2 per cent hydrocortisone cream. How much of each tube needs to be mixed? We can let x g equal the amount of cream needed from the first tube. The amount needed from the second tube is therefore $(30 - x)$ g. Then:

$$x \text{ g} \times 1\% + (30 - x) \text{ g} \times 2.5\% = 30 \text{ g} \times 2\%.$$

We can drop (i.e. cancel) each 'g' and '%' to write just

$$x \times 1 + (30 - x) \times 2.5 = 30 \times 2.$$

Therefore

$$x + 75 - 2.5x = 60$$

so that $\qquad\qquad x = (75 - 60)/(2.5 - 1) = 10.$

We conclude that 10 g are needed from the first tube and that the other 20 g should come from the second tube.

5. Estimating the composition of cells is not as straightforward as doing the same for blood plasma because allowance must be made for the presence of interstitial fluid. The amount of the latter may be found by means of a marker substance added to the bulk of extracellular fluid, *in vitro* or *in vivo*, at a known concentration. This must be something that penetrates throughout the interstitium without entering the cells or binding to their surfaces. It is allowed to diffuse into, and equilibrate with, the interstitial fluid so that the concentration there is the same as in the rest of the extracellular fluid. If the

total quantity present in a tissue sample is measured, then the amount of interstitial fluid can be calculated. The total amount of water in the sample is found by weighing it before and after drying. The calculations so far are straightforward and need not detain us. Instead, we will concentrate on the final part of the calculation and simply take it for present purposes that the following information is available: the concentration in the total tissue water, the concentration in the interstitial fluid and the fraction of the total water that is interstitial. Taking the specific example of sodium in a skeletal muscle sample, the concentration in the total tissue water is, say, 30 mmol/kg water, the concentration in the extracellular fluid is 146 mmol/kg water and the fraction of the tissue water that is interstitial is 0.15. What we need to calculate is the concentration in the cell water, which we can call x mmol/kg water.

Let us think in terms of a hypothetical 1 kg of muscle water. Its total content of sodium is 30.0 mmol. This is made up of two components. Firstly, there is 0.15 kg of interstitial water containing an amount of sodium calculated as

$$0.15 \text{ kg water} \ \times \ \frac{146 \text{ mmol}}{\text{kg water}} = 21.9 \text{ mmol.}$$

Secondly, there is $(1 - 0.15) = 0.85$ kg of cell water containing

$$0.85 \text{ kg water} \ \times \ \frac{x \text{ mmol}}{\text{kg water}} = 0.85x \text{ mmol.}$$

Thus,

$$30.0 \text{ mmol} = 21.9 \text{ mmol} + 0.85x \text{ mmol.}$$

Therefore

$$0.85x \text{ mmol} = 30.0 \text{ mmol} - 21.9 \text{ mmol} = 8.1 \text{ mmol}$$

and

$$x \text{ mmol} = \frac{8.1 \text{ mmol}}{0.85} = 9.5 \text{ mmol.}$$

The intracellular concentration of sodium is thus estimated as 9.5 mmol/kg cell water.

Several different measurements were required to make this estimate, each with the possibility of error. The estimate of the interstitial water could be particularly unreliable, as different marker substances may give different values.

6.2 What would the estimate of the intracellular sodium concentration have been if the interstitial water had been estimated to make up 20 per cent of the total?

These five rather simple examples suffice to illustrate the use of 'x' to stand for the unknown – or any other symbol one chooses. Needless to say, more complicated problems can arise that involve more elaborate algebra.

Problems

6.1 See Section 6.2.

6.2 See Section 6.2.

6.3 A flat-bottomed Petri dish has an internal radius of 4.4 cm and a depth of 1.0 cm. If it contains 30 mL of fluid, what is its depth?

6.4 500 mL of a physiological saline is made up to contain 1.2 mM inorganic phosphate. This is added as Na_2HPO_4 and NaH_2PO_4 in the ratio 4:1 (as is appropriate to pH 7.4 – see Chapter 22). What is the concentration of Na_2HPO_4?

6.5 A solution of sucrose contains 50 g per litre. What volume of this must be added to 20 mL of water to give a final concentration of 15 g/L?

6.6 Using the formula for the period of a pendulum [Eq. (1.3)], calculate the length, l, that gives a period of 2 s. Take g as 9.8 m s^{-2}.

7

Making up and diluting solutions

The preparation of solutions is a commonplace activity in many branches of biomedical science and the associated calculations require careful attention to units and their conversion factors. There are various strategies for getting the calculations right, but one particular system is explained here that can lead simply and almost automatically to right answers. The method is like that commonly used in calculating doses of medication (Chapter 8). It is not always the method of choice, but can be particularly helpful when there are many elements to a calculation. The examples here mainly concern inorganic ions, because so many of our solutions include these, but the principles apply equally to other substances.

7.1 Preparing 250 mL of 150 mM NaCl from the dry salt

A commonly prepared solution is one containing 150 mM NaCl, the simplest one-salt substitute for extracellular fluid. If 250 mL of this solution are required, how much dry NaCl needs to be weighed out? A sensible starting point is to write down the 'target' concentration in ratio form:

$$\frac{150 \text{ mmol NaCl}}{1 \text{ L}}.$$

Biomedical Calculations   Richard F. Burton
© 2008 John Wiley & Sons, Ltd

Given that the molecular mass of NaCl is 58.5, we will be using either the following relationship or its reciprocal:

$$\frac{58.5 \text{ g NaCl}}{1 \text{ mol NaCl}} = 1.$$

These expressions include both 'mmol' and 'mol', and also 'L' rather than the 'mL' of the '250 mL'. Therefore the following conversion factors will be needed (in whichever forms are appropriate):

$$\frac{1 \text{ mol}}{1000 \text{ mmol}} = 1 = \frac{1000 \text{ mmol}}{1 \text{ mol}}$$

and

$$\frac{1 \text{ L}}{1000 \text{ mL}} = 1 = \frac{1000 \text{ mL}}{1 \text{ L}}.$$

The next step is to write the ratios out as follows, together with the required volume, checking that the units can be cancelled appropriately. The answer has to work out as 'g NaCl'.

$$\frac{150 \text{ mmol NaCl}}{1 \text{ L}} \times \frac{58.5 \text{ g NaCl}}{1 \text{ mol NaCl}} \times \frac{1 \text{ mol NaCl}}{1000 \text{ mmol NaCl}} \times \frac{1 \text{ L}}{1000 \text{ mL}} \times 250 \text{ mL}$$
$$= 2.19 \text{ g NaCl}.$$

Note that, for clarity, I have chosen to include the label 'NaCl' in writing out the third ratio, as well as including it in the first two. However, there would be no ambiguity here if it were omitted, nor any uncertainty regarding the cancelling of units.

It does not matter in what order the terms are written. The above order seems a natural one, but there is also a more systematic way of choosing the sequence. According to that rationale, one records first the term in which the units in the top line are those of the final answer (g NaCl). Thereafter the units in the top lines are the units of the previous bottom lines, so that they can be cancelled immediately as each term is added to the sequence. When the uncancelled units are those of the required answer, one may be confident that every relevant factor is included in the calculation. The resulting sequence is as follows.

$$\frac{58.5 \text{ g NaCl}}{1 \text{ mol NaCl}} \times \frac{1 \text{ mol NaCl}}{1000 \text{ mmol NaCl}} \times \frac{150 \text{ mmol NaCl}}{1 \text{ L}} \times \frac{1 \text{ L}}{1000 \text{ mL}} \times 250 \text{ mL}.$$

7.2 Preparing dilutions from stock solutions

It is commonly necessary or convenient to prepare a dilute solution from one that is more concentrated. A number of dilutions of a drug may be prepared from a single more concentrated solution and, even if only one diluted solution is needed, it may be much easier and accurate to weigh out a fairly large quantity than a barely visible speck. Furthermore, it may be that the drug is being added as a concentrated solution to a known volume of physiological saline in an organ bath, in which case the appropriate volume of the added solution needs to be calculated.

When physiological saline solutions (e.g. Krebs, Ringer) have to be made up often, and perhaps with varying compositions, it is convenient to keep stock solutions of at least some of the necessary constituents (NaCl, KCl, $MgSO_4$, etc.). To minimize arithmetic, and so the danger of errors, it may also be convenient to prepare these as molar solutions. More dilute stock solutions may be appropriate for expensive substances and those of low solubility.

Suppose, for example, that one is preparing 1 L of Krebs solution in which the required potassium concentration is 4.7 mM. Potassium is generally added as its chloride. Therefore, one would simply add 4.7 mL of 1 M KCl to a 1 L flask, together with the other constituents, before making the final volume up to 1 L with distilled or deionized water. If all the salts are available as 1 M stock solutions, 1 mL made up to a litre is in each case equivalent to 1 mM. This is easy to remember and the basis may be obvious, but it is as well to spell out the relevant calculation for 4.7 mM potassium:

$$\frac{4.7 \text{ mmol KCl}}{\text{L saline}} \times \frac{1 \text{ L stock}}{1 \text{ mol KCl}} \times \frac{1 \text{ mol}}{1000 \text{ mmol}} \times \frac{1000 \text{ mL stock}}{1 \text{ L stock}}$$
$$= 4.7 \text{ mL stock/L saline.}$$

Under the previous heading we considered the preparation of 250 mL of 150 mM NaCl from *dry* NaCl. One might choose instead to prepare it from a 1 M stock solution of NaCl. To make up 1 L of 150 mM NaCl one would take 150 mL of the stock solution, so obviously one would take a quarter of that amount (= 250 mL/1000 mL) to make up just 250 mL, that is to say 37.5 mL.

You may have thought, in reading through the above calculations with their chains of multiplied ratios, that you could work out the answers more simply in your own way. So now I shall work through another example – comparing two contrasting methods. In order to accentuate the difference, the chosen example includes two extra complications. Here it is: suppose that you have a stock solution of calcium chloride ($CaCl_2$) containing 0.5 M calcium (i.e. 500 mmol/L) and that you are making up a physiological saline in a 500 mL volumetric flask that is to include 3 mequiv Ca/L. What volume of stock solution do you need to add?

One way of calculating is to reason in separate stages, like this perhaps:

1. The stock solution, 0.5 mol Ca/L, contains 0.5 mmol Ca per mL.

2. Because calcium is divalent, 3 mequiv Ca/L equates to 1.5 mmol Ca/L.

3. The 500 mL flask is to contain 0.5 L of saline and therefore 0.5×1.5 mmol Ca = 0.75 mmol Ca.

4. A 1.5 mL quantity of stock solution contains 0.75 mmol Ca.

So the answer is **1.5 mL**.

The other method involves different reasoning. First write down the relevant information, as ratios where appropriate:

$$500 \text{ mL} \qquad \frac{3 \text{ mequiv}}{1 \text{ L}} \qquad \frac{500 \text{ mmol}}{1 \text{ L}}.$$

Now reconsider each one of these individually; according to the method, what is required is that the answer should be proportional to each relevant expression, but inverted if necessary. The number of millilitres to be taken does indeed increase both with the size of the volumetric flask (500 mL) and with the required final concentration (3 mequiv/L). However, the higher the concentration of the stock solution, the less is needed. Therefore the third item needs to be inverted and written as

$$\frac{1 \text{ L}}{500 \text{ mmol}}.$$

The required set of expressions is therefore amended to:

$$500 \text{ mL} \qquad \frac{3 \text{ mequiv}}{1 \text{ L}} \qquad \frac{1 \text{ L}}{500 \text{ mmol}}.$$

Some conversion factors may be needed too, written as ratios that each equal 1:

$$\frac{1 \text{ mmol}}{2 \text{ mequiv}} \qquad \frac{1000 \text{ mL}}{1 \text{ L}} \qquad \frac{1000 \text{ mmol}}{1 \text{ mol}} \qquad \left(\text{and possibly } \frac{58.5 \text{ g}}{1 \text{ mol}} \right).$$

The task now is to combine all of the first three expressions and whichever of the conversion factors prove(s) appropriate. Bear in mind that the conversion factors may need to be used as their reciprocals. The following calculation contains just one conversion factor.

$$? \text{ mL} = 500 \text{ mL} \times \frac{3 \text{ mequiv}}{1 \text{ L}} \times \frac{1 \text{ L}}{500 \text{ mmol}} \times \frac{1 \text{ mmol}}{2 \text{ mequiv}} = \textbf{1.5 mL}.$$

This calculation follows no special sequence, except that the first three terms are in the random order given above and that a necessary conversion factor is added at the end. [If, by the way, an extra conversion ratio were written down, such as (58.5 g)/(1 mol), any attempt to use it would be automatically thwarted.]

   This works well, but it is also possible to follow instead the rationale given above in relation to the solution of sodium chloride. According to this, the first item is again '500 mL', but chosen now because it matches the unit of the final answer. Thereafter, each ratio includes a unit in the numerator or denominator that cancels with the identical unit in the preceding ratio, until mL, having temporarily disappeared through cancellation, reappears to be left on its own in the answer. Thus,

$$? \text{ mL} = 500 \text{ mL} \times \frac{1 \text{ L}}{1000 \text{ mL}} \times \frac{3 \text{ mequiv}}{\text{L}} \times \frac{1 \text{ mmol}}{2 \text{ mequiv}} \times \frac{1 \text{ L}}{500 \text{ mmol}} \times \frac{1000 \text{ mL}}{1 \text{ L}}$$
$$= \textbf{1.5 mL}.$$

Using the conversion factor for mL and L twice may look superfluous, but it does allow the sequence to progress appropriately, on auto-pilot so to speak, and with minimum thought. Which method do *you* prefer? You may choose to combine elements of both types of calculation. You are also free to do any unit conversions in your head if they come easily and reliably. For example, you may see '500 mL' automatically as 0.5 L and see no point in writing

$$500 \text{ mL} \times \frac{1 \text{ L}}{1000 \text{ mL}}.$$

   If you want firmer advice on the best method, I would merely suggest that you fully understand the final and most orderly one, before perhaps rejecting it. It is a way of calculating that works in other situations too.

   How foolproof is the latter method? Suppose that one of the initial ratios is mistakenly inverted. The result would not be a wrong answer, but a failure of the cancelling of units to work out sensibly and so give an answer at all. With both initial ratios inverted (3 mequiv/L and 1 L/500 mmol), the units would work out correctly, but the answer would be an unlikely one (166 667 mL). Thinking about whether an answer is a likely one is a test that some students overlook.

Problems

7.1 The molecular masses of Mg, $MgSO_4$ and $MgSO_4$. $7H_2O$ are respectively 24.3, 120.4 and 246.5. How much of the hydrated salt, $MgSO_4$. $7H_2O$, is needed to make up a 1 M stock solution of $MgSO_4$ in a 250 mL volumetric flask?

7.2 A 0.3 mL quantity of a solution containing substance A at a concentration of 50 g/L is added to an organ bath holding 30 mL of physiological saline. What is the resulting concentration of A in the organ bath?

7.3 A piece of guinea pig ileum is set up in an organ bath containing 25 mL of physiological saline. Acetylcholine is available as a 2.5×10^{-6} M solution. What volume of that should be added to the bath to obtain a concentration of 3×10^{-8} M?

7.4 You have 250 mL of a 1 M stock solution of sodium bicarbonate ($NaHCO_3$). A physiological saline is being made up in a 1 L volumetric flask to include bicarbonate at 25 mequiv/L. What volume of stock solution needs to be added? The molecular mass of $NaHCO_3$ is 84.01.

7.5 The molar mass of sucrose is 342.3, so that 1 mmol $= 342.3$ mg. What mass of sucrose is needed to make up a 200 mM solution in a 100 mL volumetric flask?

7.6 You have a stock solution of glucose with a concentration of 90 g/L. A physiological saline is to be made up in a 250 mL volumetric flask and is to include glucose at 5 mmol/L. What volume of stock solution needs to be added? The molecular mass of glucose is 180.

8.

Calculating drug doses

Calculating drug doses is routine in medicine, but mistakes can be fatal. Various standard methods of calculation are taught, but one that is much advocated is entirely in keeping with the principles and procedures of unit analysis as already discussed. Other names are used for it, however, such as 'factor-label method', 'label-factor method' and 'unit factor method'. Also much used is the name 'dimensional analysis', which is a pity, since that name has long been applied to a different, though related, system of analysis (see Chapter 1).

Here is a simple problem. A patient weighs 65 kg and is to be given 3 mg of a drug for every kilogram of body mass. What is the correct dose? Although the calculation is easy, it is written out here in a way that makes obvious the need to cancel 'kg':

$$\frac{3 \text{ mg}}{\text{kg}} \times 65 \text{ kg} = 195 \text{ mg}.$$

Calculation may involve more steps. One may solve each individual problem by applying common-sense reasoning, but it is clearly desirable to have a standard method of calculation that should lead infallibly to the right answer. Suppose that a 70 kg man is prescribed aminophylline as an infusion at a dose of 0.5 mg/kg per hour (0.5 mg kg^{-1} h^{-1}) and that the drug is available in 10 mL ampoules that each contain 250 mg. How many millilitres would be needed if the infusion is to

Biomedical Calculations   Richard F. Burton
© 2008 John Wiley & Sons, Ltd

last 12 h? Let us write down the information, in no particular order:

$$70 \text{ kg} \qquad \frac{0.5 \text{ mg}}{\text{kg h}} \qquad \frac{250 \text{ mg}}{10 \text{ mL}} \qquad 12 \text{ h}$$

Can we just multiply them all together? The answer has to be in millilitres and it is the third item that includes 'mL'. There, however, the mL is in the bottom line, so either division is called for, or the expression needs to be written as its reciprocal (which amounts to the same thing). The latter is preferable in that the subsequent treatment of units is more transparent. So let us see if the following calculation works:

$$70 \text{ kg} \times \frac{0.5 \text{ mg}}{\text{kg h}} \times \frac{10 \text{ mL}}{250 \text{ mg}} \times 12 \text{ h}.$$

Looking just at the units, one sees that 'kg' appears once in a numerator (top line) and once in a denominator (bottom line) and so would cancel. The same is true of 'mg' and 'h'. Therefore the only unit to survive the multiplication and cancellation would be 'mL' and that is the correct outcome. All that remains to do, therefore, is the arithmetic – obtaining an answer of 16.8 mL.

The calculation could have been more complicated – if, for example, the patient's body mass were known only in pounds and the amounts and concentrations of drug were differently expressed in terms of mg, μg or μmol.

The four terms in the above calculation are presented in no particular sequence, nor does the order strictly matter. However, a more rational sequence is sometimes recommended (see Notes). Here it is.

First identify what needs to be calculated and write it down as the left side of an equation that is to be built up. It should generally be expressed as a fraction, such as '? mL/dose' or 'x mL/dose'. The right side generally consists of a succession of ratios to be multiplied, of which the first has in its numerator the 'label' (i.e. unit) of the final answer (as recorded on the left side of the equation). In the above example, the appropriate ratio would be '10 mL/250 mg'. The unit of the denominator will not be appropriate in the final answer, so the next ratio should be one that contains it in the numerator, allowing immediate cancelling. One continues in this way so that undesired denominator units are, if possible, consecutively cancelled by succeeding undesired numerator units. The process stops automatically when the uncancelled denominator label is as required and no unwanted unit remains. That probably seems abstract and confusing, so here is the above calculation in that form:

$$\frac{?\,\mathrm{mL}}{\mathrm{infusion}} = \frac{10\,\mathrm{mL}}{250\,\mathrm{mg}} \times \frac{0.5\,\mathrm{mg}}{\mathrm{kg\,h}} \times \frac{12\,\mathrm{h}}{\mathrm{infusion}} \times 70\,\mathrm{kg}$$

$$= \frac{16.8\,\mathrm{mL}}{\mathrm{infusion}}.$$

Look at this and work through the sequence of cancellations. In this particular case the last two terms (12 h/infusion and 70 kg) could be in the reverse order, as both 'kg' and 'h' are in the previous denominator. You will have spotted that the final term, '70 kg' is not a ratio. If you wish, you can express it as a ratio by dividing it by one and writing '70 kg/1'. Alternatively, you could write '70 kg/patient', in which case 'patient' would need to be included in the denominator on the left side of the equation. There seems little point to that, but it is a matter of personal choice.

Note that this method of calculating doses allows for a neat and systematic organization of the calculation rather than a disorganized series of steps. It also eliminates the need for a multiplicity of formulae to be separately remembered. Furthermore it prevents any accidental inversions of individual ratio terms. Much less reasoning is needed than if each situation has to be thought through from scratch, although it would be rash to abandon critical thinking altogether. In particular, one should always ask oneself whether the calculated dosage seems plausible.

When it comes to details, some further thought may be needed. Thus, in a different situation, it should be recognized that '75 mg tablets' implies the ratio form '75 mg/tablet'. The expression '5 per cent NaCl (w/v)' would need to be interpreted as '5 g/100 mL' (although it could enter a calculation as its reciprocal '100 mL/5 g').

Problems

8.1 Polycillin is available as 250 mg in 5 mL of solution and a dose of 0.50 g is to be administered every 12 h. How many millilitres would be given each time?

8.2 The dose of a particular drug preparation is specified as 200 minims. What is that in millilitres? Take 1 mL as 16.9 British ('imperial') minims or 16.2 US minims.

8.3 A dose of 0.15 grains of morphine is prescribed. It is available as 10 mg tablets. How many tablets would be required? 1 grain = 64.8 mg.

8.4 The dose of a drug is specified as 300 mg per square metre of body surface area (300 mg/m^2). A patient's surface area is calculated as being 1.69 m^2 (see Section 11.17). What is the appropriate dose?

8.5 A patient weighing 68 kg is to be infused with sodium nitroprusside at a rate of 3 μg per kg of body mass per minute. An infusion pump is set up to contain 50 mg of sodium nitroprusside in 50 mL of solution. What is the appropriate pump rate? The answer should be expressed in mL/min. It is to be calculated from the following quantities:

$$68 \text{ kg} \qquad \frac{3 \text{ μg}}{\text{kg min}} \qquad \frac{50 \text{ mg}}{50 \text{ mL}}$$

Set out the calculation in the recommended manner, adding any necessary conversion factors.

8.6 Trimethoprim is available as a suspension of 50 mg per 5 mL. A child of 24 kg is to be given a dose of 4 mg/kg. What volume is required?

8.7 Cefaclor is to be given three times a day to a 150 lb patient. The required dose is 30 mg per kilogram body mass per day. The antibiotic is available as powder yielding 250 mg/5 mL when reconstituted. What volume should be administered each time?

9

More about solutions – electroneutrality, osmotic pressure and activity

9.1 The principle of electroneutrality

The principle of electroneutrality (or principle of electrochemical neutrality) is one of the most important fundamental quantitative ideas in relation to electrolyte regulation, including acid–base balance and aspects of renal function, although, curiously, many books on these topics fail to say so. According to this principle the total amount or concentration of positive charges in a solution is virtually equal to the total amount or concentration of negative charges. This is where equivalents come into their own, as opposed to moles, for another way of expressing the principles is as in this box:

Total amounts, or concentrations, of anions and cations are virtually equal when expressed as equivalents or milliequivalents.

There are in fact minute discrepancies associated with the existence of cell membrane potentials, but these are too small to be demonstrated by chemical analysis. This is why the statement includes the word 'virtually'. We return to this point below.

Biomedical Calculations Richard F. Burton
© 2008 John Wiley & Sons, Ltd

The principle has important implications. For example, the excretion of acid in urine is quantitatively linked to the excretion of ions such as sodium, potassium and chloride. The concentration of bicarbonate in blood plasma, so important to pH regulation, cannot change without concomitant changes in other ions present, including, for example, buffer ions. Except in special circumstances (see below), charges balance as ions move across epithelia, sometimes with anions and cations moving in the same direction, sometimes with one kind of ion exchanging with another of like charge (e.g. chloride exchanging with bicarbonate across the erythrocyte membrane). That there are these links amongst the concentrations and movements of ions follows from the principle of electroneutrality. It is a separate issue that there may also be specific mechanisms involved, including the variety of active and passive ion transport mechanisms located in cell membranes. Some are electroneutral, like sodium/proton exchangers. Others do transport a net charge, like the sodium–potassium pump, but other ion movements must occur that maintain electroneutrality.

Consider now the composition and preparation of simple salt solutions, such as physiological salines. The constituent ions (e.g. Na^+, K^+, Ca^{2+}, Cl^-, HCO_3^-) cannot be added to a solution individually, but are weighed out as salts (e.g. NaCl, $CaCl_2$). (Dropping in metallic sodium may be fun, but both anions and cations form during the reaction.) For each salt the numbers of equivalents of cation and anion are necessarily equal so that the principle of electroneutrality necessarily applies to the final solution, though not only for that reason. Suppose that one needs a solution with the composition shown in Table 9.1. For this solution to have the pH of extracellular fluid (with reduced likelihood of the otherwise high pH causing calcium carbonate to precipitate), it would need to be gassed with carbon dioxide (approximately 5 per cent), but that does not concern us here. Magnesium is commonly added as sulphate ($MgSO_4.7H_2O$) because magnesium chloride is hygroscopic.

9.1   What other salts you would use in making up this solution, and at what concentrations?

Table 9.1   Cation and anion concentrations in the required solution

Cations		Anions	
Na^+	145 mM	Cl^-	129 mM
K^+	5 mM	HCO_3^-	25 mM
Mg^{2+}	1 mM	SO_4^{2-}	1 mM
Ca^{2+}	2 mM		

Table 9.2 Here the concentrations of Table 9.1 are expressed in mequiv/L

Cations		Anions	
Na^+	$145 \times 1 = 145$ mequiv/L	Cl^-	$129 \times 1 = 129$ mequiv/L
K^+	$5 \times 1 = 5$ mequiv/L	HCO_3^-	$25 \times 1 = 25$ mequiv/L
Mg^{2+}	$1 \times 2 = 2$ mequiv/L	SO_4^{2-}	$1 \times 2 = 2$ mequiv/L
Ca^{2+}	$2 \times 2 = 4$ mequiv/L		
Totals	156 mequiv/L		156 mequiv/L

The tabulated concentrations may be re-expressed in terms of mequiv/L by multiplying them by the respective valencies to produce Table 9.2. When such solutions are formulated, it may be worth checking the balance of anions and cations in this way, especially when the solutions are more complicated or esoteric, or when someone else's formula is being adopted. Published formulae are occasionally wrong, sometimes as a result of copying errors, one presumes, but researchers who *use* physiological salines tend in general to be less interested in their composition and rationale.

Inorganic phosphate ions have not been mentioned so far, although they occur at a concentration of about 1 mM in extracellular fluid and at higher concentrations when used for buffering artificial solutions. At a physiological extracellular pH of 7.4 inorganic phosphate takes two main forms, namely HPO_4^{2-} and $H_2PO_4^-$, with these being in the ratio 4:1. Unionized H_3PO_4 and fully ionized PO_4^{3-} exist in only trace amounts at that pH. In the context of electroneutrality calculations and the checking of anion–cation balances, one may treat inorganic phosphate as a single component having an appropriate mean valency and a known total concentration. This is easier than first calculating the individual concentrations of HPO_4^{2-} and $H_2PO_4^-$. The mean valency is obviously nearer 2 than 1 and is in fact 1.8 at pH 7.4. This is calculated as $(2 \times 4 + 1)/(4 + 1)$. Urine contains much higher concentrations of inorganic phosphate and the ratio of HPO_4^{2-} to $H_2PO_4^-$ in urine needs to be calculated from the (very variable) pH (Chapter 22). (Phosphate is an important urinary buffer, with HPO_4^{2-} taking up secreted hydrogen ions to become $H_2PO_4^-$.)

Hydrogen ions have not been explicitly mentioned until the last sentence, despite their importance. This is because their concentrations in cells and extracellular fluid are minute compared with those of other inorganic ions (Chapter 21). This is true of hydroxyl ions also, so they only contribute significantly to anion–cation balance at extremes of pH.

9.2 But what about membrane potentials and short-circuit currents?

We now come to 'bioelectricity', but only to make a few simple points. This is another context in which ionic charges are significant. However, ionic movements are emphatically not electroneutral in the context of action potentials in nerve and muscle. The depolarization and reversal of membrane potential during an action potential involves a net inward movement of positive ionic charges (carried, in the best known case, by sodium ions). This departure from electroneutrality constitutes an electric current. The resting membrane potential of a cell is likewise due to a departure from electroneutrality, since it results from very slight differences between the equivalent concentrations of anions and cations. There is a slight preponderance of anions (with their negative charges) inside the cell. The extent of the discrepancy depends on such things as the cell size and magnitudes of the membrane potential, but can be, say, 10^{-3} to 10^{-2} mmol L^{-1}. Such tiny discrepancies are far too small to be demonstrated by chemical analysis, and that is why I wrote above that the total amounts or concentrations of cations and of anions, in equivalents, are *virtually* equal. This paragraph is in one sense just a footnote to what comes before it, but it is worth emphasizing the smallness of the concentration changes associated with action potentials, because a belief in big changes is too common a misconception.

Small change from a millionaire may save a beggar with virtually no effect on the rich person's wealth. Likewise, membrane potentials and hydrogen and hydroxyl ions near neutrality, despite their importance, are irrelevant to most arithmetic of electroneutrality.

It must also be acknowledged in passing that there are special experimental contexts in which large movements of either anion or cation occur that are *not* balanced by movements of the other. This situation arises in studies of ion transport when the movement of one kind of ion is being balanced by an electric current applied from an external source through electrodes. (For example, this so-called 'short-circuit current' is a convenient way of studying sodium transport across sheets of frog skin.)

9.3 Anion gap

The principle of electroneutrality may be applied to extract extra information from a limited knowledge of the ionic composition of a particular fluid. Routine clinical analyses of blood plasma are inevitably incomplete, but one can infer something about unmeasured ions from the ion concentrations that are known. In terms of milliequivalent concentrations:

$$[\text{all measured cations}] - [\text{all measured anions}]$$
$$= [\text{all unmeasured anions}] - [\text{all unmeasured cations}].$$

Most often the unmeasured anions exceed the unmeasured cations and then the difference between them can be called the 'anion gap' or 'anion deficit'. In principle, the more ions there are involved in the calculation of the anion gap, the more informative it is about the unmeasured ions. However, some ions contribute too little to make much difference (e.g. Mg^{2+} and various organic ions) and, to minimize the number of necessary chemical analyses, the number of ionic species involved in the calculation should be kept small. Two definitions of anion gap are in general use that use minimal information:

$$[Na^+] - ([Cl^-] + [HCO_3^-])$$

and

$$([Na^+] + [K^+]) - ([Cl^-] + [HCO_3^-]).$$

These use concentrations in mequiv/L, which, for these univalent ions, are numerically the same as concentrations in mmol/L. The anion gap needs to be expressed in mequiv/L because the unmeasured anions are unlikely all to be univalent. Most protein molecules carry a large net negative charge.

The normal range for the anion gap of plasma calculated without using potassium is commonly said to be 8–16 mequiv/L. A high anion gap may have a variety of causes, most of which lower the bicarbonate concentration. These include the presence of β-hydroxybutyrate in ketoacidosis, renal insufficiency with retention of sulphate or phosphate, lactic acidosis, ingestion of salicylate, oxalate produced from ingested ethylene glycol and formate produced from methanol.

Another version of the anion gap is used in the context of acid-base physiology (Chapter 23), namely the 'strong ion difference' or SID. This may be defined as the total concentration of strong cations minus the total concentration of strong anions (both in terms of milliequivalents). The 'strong ions' are the non-buffer ions such as Na^+ and SO_4^{2-} that are not directly affected by pH, unlike HCO_3^-, $H_2PO_4^-$ and HPO_4^{2-}. However, they may be taken to include buffer ions, such as lactate, that are unaffected by pH in the physiological pH range. The SID thus equals the net anionic contribution of bicarbonate and buffers such as plasma proteins and inorganic phosphate. Conceptually, the SID so defined is useful in emphasizing the determinants of plasma bicarbonate concentration (Chapter 23). Practically, it has the disadvantage of requiring too many separate analyses to be useful. For this reason the SID is sometimes defined much more simply as $([Na^+] + [K^+] - [Cl^-] - [lactate^-])$ or just $([Na^+] + [K^+] - [Cl^-])$.

The strong ion difference, as $([Na^+] + [K^+] - [Cl^-])$, is also used for urine under the name of 'urine anion gap'. It becomes more negative, for example, when extra chloride ions are excreted with ammonium ions. This relates to 'titratable acidity', which is a measure of urinary acid excretion (see Chapter 23).

9.4 Osmoles and osmolality

Osmotic pressure is important for its role in the movement of water within the body – across cell membranes generally, and in large amounts from the gastrointestinal tract and renal tubules. The osmotic pressure of a solution is determined by the total concentration of all the solutes present, both ions and non-electrolytes (see Notes). In animal physiology, but not plant physiology, one is almost entirely concerned with the concentrations rather than the actual osmotic pressures. (Colloid osmotic pressures are another matter – we come to them in Section 14.4.) In this context the concentrations of each solute are expressed in special units, namely *osmoles* or *milliosmoles*. In an 'ideal', or 'infinitely dilute', solution, an osmole equals a mole, but this is only roughly true in body fluids. (Nevertheless, some textbooks do incorrectly equate osmoles and moles.) For accurate work it is therefore necessary to apply correction factors, known as *osmotic coefficients*. Thus, for any particular solute,

$$1 \text{ milliosmole} = 1 \text{ millimole} \times \text{ relevant osmotic coefficient.}$$

Take, for example, the case of a solution of sodium chloride at a concentration of 150 mmol/kg water at 25 °C. The total concentration of ions, sodium and chloride, is $2 \times 150 = 300$ mmol/kg water. In this case, published tables tell us that the osmotic coefficient is 0.928. Therefore the concentration is $300 \times 0.928 = 278$ milliosmoles/kg water, abbreviated as 278 mosmol/kg water. The osmotic coefficient, often denoted ϕ, is usually given without units, but these may be taken as osmol/mol or mosmol/mmol.

Concentrations expressed this way are known as 'osmolalities'. There are also the words *osmolal*, *osmolarity* and *osmolar* that correspond to molal, molarity and molar. It is better to use osmolalities rather than osmolarities, because they reflect osmotic properties better and are not influenced by the presence of osmotically inactive substances, e.g. lipids, that may contribute to volume.

Osmotic coefficients for any given substance are not constant. Thus ϕ for NaCl is exactly 1 in very dilute solutions, but falls as the concentration increases up to 0.4 molal NaCl. At that concentration ϕ reaches a minimum value of 0.920. Coefficients are also affected by other solutes present. For sucrose on its own ϕ is slightly above 1. These details need not concern most of us, but it is possible to look up osmotic coefficients for particular substances. There is obvious potential for complications and uncertainties here, but there is also the possibility of direct measurement. Accurate measurement of osmotic pressure itself is difficult, but there are two related properties that can be accurately measured with very small samples, namely freezing point depression and vapour pressure (see Notes). These are proportional to osmolality and are what commercial 'osmometers' measure.

The osmolality of most human body fluids is regulated close to 300 mosmol/kg water. That is the round-number value that is usually to be found in textbook diagrams of the renal countercurrent system. Actual values are typically 285–295 mosmol/kg water. Most cells have the same osmotic pressure as the extracellular fluid. Cells and extracellular fluid are accordingly described as 'isosmotic'. The word isosmotic is also applied to artificial saline solutions made up to have the same osmotic pressure as tissues. Cells with rather different osmolalities include epithelial cells in contact with fluids that are not isosmotic with the body fluids in general, notably in the renal medulla.

'Isotonic' is often used to mean 'isosmotic', but strictly it means something different. A solution is isotonic to particular cells if contact between them does not cause the cells to change in volume as a result of osmosis. Erythrocytes suspended in an isosmotic solution of urea swell and almost immediately haemolyse, so that the solution is not isotonic. This is because urea rapidly diffuses into the erythrocytes, thus raising their internal osmotic pressure and drawing in water until the stretched cell membranes become leaky and release their contents. (The term 'isotonic' was first used in relation to plant cells, which are prevented from excessive swelling by their cellulose cell walls.)

Let us try a rough calculation of the osmolality of plasma using the representative and approximate concentrations of its main constituents shown in Table 9.3. The anions and cations do not balance in terms of charge, mainly because plasma proteins are not included. Despite their bulk and high net negative charge, proteins make a trivial contribution to the osmolality (see Section 14.4). The sum of all these concentrations is 304 mmol/kg water. The contributions of other plasma solutes, such as amino acids, are usually fairly small. The osmotic coefficients vary from substance to substance, but, in the context of plasma variability, we will not

Table 9.3　Representative concentrations of solutes in blood plasma.

Solute	Concentration (mmol/kg water)
Sodium	150
Chloride	110
Bicarbonate	25
Potassium	5
Free calcium	1
Free magnesium	1
Inorganic phosphate	1
Urea	5
Glucose	5
Lactate	1
Total	304

be far wrong if we apply an overall value of 0.93 mosmol/mmol, rounded up from the value given above for NaCl. Then:

$$304 \text{ mmol/kg water} \times 0.93 \text{ mosmol/mmol} = 283 \text{ mosmol/kg water.}$$

This is just under the range given above for typical plasma (285–295 mosmol/kg water), but remember that the small amounts of other solute have been neglected.

9.5　Osmolar gap

Given a solution with exactly the composition shown in Table 9.3, one could check its actual osmolality with an osmometer. If the solution was actually a plasma or serum sample and the actual osmolality turned out to be significantly higher than calculated, one might presume that some other substance was present that had not been measured. This could be a clue to the presence of something clinically important, such as methanol, ethylene glycol, paraldehyde or lithium. This brings us to the idea of the 'osmolar gap'. This is the difference between the measured and calculated osmolalities. (Other names for this include osmole gap, osmolal gap and osmolarity gap.)

Most of the total plasma osmolality is due just to a few solutes. Therefore, for defining and calculating the osmolar gap, one may use just those few in estimating the osmolality. Normal variations in the rest should be unimportant. The question now is which solutes to choose. Recall that, in terms of milliequivalents, the total concentration of anions equals the total concentration of cations. Therefore, even taking the sodium concentration on its own might not be useless, if doubled, because doubling would roughly take care of accompanying anions. Actually, a number of better formulae have been proposed for estimating the osmolality. Here is one of them:

$$\text{plasma osmolality} \approx 2([\text{Na}] + [\text{K}]) + [\text{urea}] + [\text{glucose}]. \qquad (9.1)$$

Here the individual solute concentrations are in mmol/L, which is how they are usually measured. That may seem inappropriate, but in fact it is not. For a typical plasma protein concentration or plasma water content, the sum can be converted to mmol/kg water by dividing it by 0.94 (Section 5.3). Then one can convert the millimoles to milliosmoles by multiplying the result by 0.93 mosmol/mmol (as above). The ratio 0.93/0.94 (= 0.99) is so near 1 that these two correction factors virtually cancel each other out and can be ignored. Thus the estimate, based on only four analyses, is actually quite good for practical purposes. In this case the osmolar gap is given by the discrepancy between the actual and calculated osmolality:

$$\text{plasma osmolality} - \{2([\text{Na}] + [\text{K}]) + [\text{urea}] + [\text{glucose}]\}.$$

There are variants on Eq. (9.1). The potassium term may be dropped on the grounds that the concentration is fairly small and varies too little to matter much. The urea term may be replaced by an equivalent term for blood urea nitrogen (BUN). The glucose term is sometimes replaced with the glucose concentration in mg/dL divided by 18 (one-tenth of the molar mass). One variant that has been shown to work well is this:

$$\text{plasma osmolality} \approx 1.86 \, [\text{Na, mM}] + [\text{BUN, mg/dL}]/2.8$$
$$+ \, [\text{glucose, mg/dL}]/18 + 9. \qquad (9.2)$$

The '+ 9' is included to produce an osmolar gap that is normally near zero.

I give two formulae for several reasons: firstly to illustrate both the principles and the arbitrariness of the choice and, secondly, as a warning that one may need to know what formula is being used in a particular hospital before interpreting the figures. The normal range of osmolar gaps will depend on the formula used. Values may be positive or negative.

> **9.2** In a real case study (Sood and Richardson, 2007), a man presented in a state of clouded consciousness. He had been receiving lithium therapy. Plasma sodium was 142 mmol/L, glucose was 4.9 mmol/L, urea was 6.6 mmol/L. His osmolality was calculated from these molar concentrations as $(2[Na] + [glucose] + [urea])$. His measured plasma osmolality was 324 mosmol/kg water. What was his osmolar gap as calculated from these data? His plasma contained lithium (Li^+) at 14.5 mmol/L. How does that relate to the osmolar gap?

9.6 Osmosity

Osmotic concentrations may also be expressed as 'osmosities'. The 'osmosity' of a solution is the molar or millimolar concentration of a pure sodium chloride solution of the same osmotic pressure. This is practically convenient in that osmometers may be calibrated directly against standard solutions of sodium chloride. There is then no need to consider osmotic coefficients.

What is a typical blood osmosity? We have to consider a solution of sodium chloride having the osmolality of typical plasma, namely 285–295 mosmol/kg water. If we take the osmotic coefficient as 0.93 mosmol/mmol and take 'sodium chloride' as separate sodium and chloride ions, then the concentration of sodium chloride is

$$285\text{–}295 \text{ mosmol/kg water} \div 2 \div 0.93 \text{ mosmol/mmol}$$

$$= 153\text{–}159 \text{ mmol/kg water.}$$

For protein-free salt solutions it was established in Chapter 5 that concentrations are numerically very similar whether expressed in terms of volume or of mass of water (molar or molal). We can therefore say that the osmosity of plasma is typically about 153–159 mmol/L.

9.7 Cell contents

One of the many ways that cells differ from blood plasma is their high content of protein and other materials. These occupy space and accentuate the difference between molar and molal concentrations. Actually, molar concentrations are rarely considered, because amounts of tissue and cell are almost always determined in terms of mass, not volume. Concentrations are therefore expressed either for cell water or for total cell mass. To take the convenient example of the erythrocyte, its water content is about 72 per cent by mass. Therefore 1 kg of cell water corresponds to about 1.4 kg (= 100/72 kg) of red cells. Concentrations expressed per kg of water and per kg of cells differ in the same ratio, the latter being numerically the lower.

Unlike concentrations in plasma, intracellular concentrations are not easily and routinely measured. This is mainly because tissues contain not only cells, but also interstitial fluid. The composition of this may be taken as being like that of other more accessible extracellular fluid (e.g. blood plasma), but the amount of interstitial fluid in a tissue sample needs to be estimated before the composition of the cells can be worked out (see Chapter 6). Tissue analyses are thus a matter for specialist researchers, whereas the typical reader is concerned with generalizations like the following.

Most cells are isosmotic with blood plasma. They contain much higher concentrations of potassium and low concentrations of sodium and chloride. The large amount of protein present contributes rather little to osmolality, but has a large net negative charge. A variety of other organic substances are present, these being mostly nitrogen-containing compounds with no net charge. These include amino acids.

Although osmosity is properly defined in terms of molar or millimolar concentrations, the concept can be adapted to millimolal concentrations. There are abundant published data on intracellular concentrations of sodium, potassium and chloride that are coupled with information on extracellular osmolality. If the latter is taken as equal to intracellular osmolality, it can be expressed as intracellular osmosity, namely the equivalent concentration of NaCl in mmol/kg water. It is therefore possible to calculate what one might call an intracellular 'osmosity gap' that is somewhat analogous to the osmolar gap, though it has fewer terms:

$$2 \times \text{osmosity} - [K] - [Na] - [Cl].$$

As an example, suppose that a sample of mammalian skeletal muscle has an osmosity of 155 mmol/kg water, determined for extracellular fluid on the assumption that it is isosmotic, and that the cells contain potassium, sodium and chloride at the following concentrations:

K	168 mmol/kg water
Na	25 mmol/kg water
Cl	7 mmol/kg water.

The osmosity gap is $(2 \times 155 - 168 - 25 - 7) = 110$ mmol/kg water. Because the proteins contribute little to the osmotic pressure, the osmosity gap is largely due to small organic molecules. The osmosity gap is thus an indirect measure of the total concentration of such substances. For many tissues, these have not all been directly quantified.

The osmosity gap may be compared with a corresponding anion gap, defining the latter here as ([K] + [Na] − [Cl]). In this case the anion gap is (168 + 25 − 7) = 186 mequiv/kg water. This is largely due to the net negative charge on intracellular proteins (which contribute hardly at all to the osmosity gap). What the calculation of these osmosity and anion gaps illustrates is the possibility of quantifying something by actually measuring something else (see Notes).

9.3 A sample of muscle from a crab has an osmosity of 500 mmol/kg water – high like that of the sea water it lives in. The muscle contains potassium, sodium and chloride at the following intracellular concentration:

$$
\begin{aligned}
&\text{K} \quad 149 \text{ mmol/kg water}\\
&\text{Na} \quad 40 \text{ mmol/kg water}\\
&\text{Cl} \quad 46 \text{ mmol/kg water.}
\end{aligned}
$$

Calculate the osmosity gap and anion gap. The cells of marine invertebrates often contain large amounts of solute with no net charge (many of the compounds having one negative and one positive charge). Comparing the crab data with those for the mammalian muscle, consider whether that might be true here.

9.8 Effective osmolality, effective osmotic pressure

The terms osmolality, osmosity and osmotic pressure refer strictly to physical properties of solutions that can be measured by an osmometer. However, what governs the movements of water between cells and extracellular fluid is the 'effective osmotic pressure'. Any dissolved substance that distributes itself evenly across cell membranes by simple diffusion cannot, at equilibrium, have any osmotic effect on the distribution of water within the body. Such a substance is urea. Inasmuch as this therefore makes no effective contribution to osmotic pressure, its contribution to the total osmolality may be subtracted to give the 'effective osmolality'. For typical, normal concentrations of urea, this means that plasma osmolality is effectively reduced by about 5 mosmol/kg water. That makes a difference of less than 2 per cent, so is not of great importance, but the difference can be considerable in uraemia, with urea concentrations rising even above 100 mosmol/kg water. Sharks, incidentally, are notable for the very high concentrations of urea in their body fluids, exceeding 300 mM, for example. This contributes to the very high total osmotic pressure of their intracellular and extracellular body fluids, which approximates to that of sea water.

9.9 Osmotic shifts of water between cells and extracellular fluid

Most cells are in osmotic equilibrium with the extracellular fluid, so that the osmotic pressure is essentially uniform throughout the body. For present purposes we will ignore the gradients in limited regions such as the renal medulla. The status quo may be temporarily upset, as when water is absorbed after a drink and dilutes first the extracellular fluid and then the intracellular fluid. At other times shifts of solute between the two compartments lead to accompanying shifts of water. The osmotic pressure in both compartments may also alter together as solutes are gained or lost by the body. Let us explore such effects in terms of a simple two-compartment model of the body. Secondary effects, as on renal excretion, are ignored.

Initially the intracellular and extracellular compartments are postulated as containing 28 and 14 kg of water, respectively. (These are typical textbook values, although those are usually given in litres. They accord with a round 60 per cent water for a standard 70 kg man.) The initial osmolality is taken as 285 mosmol/kg water (this being the effective osmolality and not including urea). From these figures it follows that the initial solute contents of the intracellular and extracellular compartments are respectively 7980 and 3990 mosmol (i.e. 28 and 14 kg water, each multiplied by 285 mosmol/kg water). Figure 9.1 shows the body's initial

intracellular	extracellular
water:	
28 kg	14 kg
solute:	
7980 mosmol	3990 mosmol
concentration:	
285 mosmol /kg water	285 mosmol /kg water

Figure 9.1 The two rectangles represent intracellular and extracellular water in a model body. Within them are shown the respective masses of water and also the amounts and concentrations of solute.

condition. Responses other than osmotic water shifts, including renal excretion, are not considered here.

The simplest change to consider is the infusion of isosmotic fluid into the extracellular fluid. The osmotic equilibrium is not upset and the cell water is unaffected. Conversely, diarrhoea with loss of isosmotic fluid causes depletion just of the extracellular fluid, with no change in osmolality.

When water is absorbed following a drink, the extracellular fluid increases in volume and its osmotic pressure is lowered. This causes water to be taken up by the cells until osmotic equilibrium is restored. To illustrate the effects, let us take the mass of extra water as 1 kg. This raises the total body water from 42 to 43 kg. Once the two body compartments have reached a steady state, the osmolality is the same in both. It is easily calculated by dividing the total number of osmoles in the body (7980 + 3990 = 11 970) by the new total water content of 43 kg. The result is 278.4 mosmol/kg water. The extra 1 kg of water is shared between compartments in proportion to their initial sizes. Thus the cells gain 0.67 kg (1 kg × 28 kg/ 42 kg) and the extracellular fluid gains 0.33 kg (1 kg × 14 kg/42 kg). As a check, note that the final volumes can also be calculated respectively as

$$\frac{7980 \text{ mosmol}}{278.4 \text{ mosmol/kg water}} = 28.7 \text{ kg water}$$

and

$$\frac{3990 \text{ mosmol}}{278.4 \text{ mosmol/kg water}} = 14.3 \text{ kg water}.$$

The final result is shown in Figure 9.2.

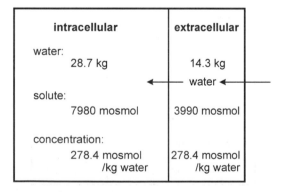

Figure 9.2 This new version of the model body shows the effects of adding 1 kg of water (the result of drinking). Following equilibration, the water is distributed in both compartments and the osmolal concentrations are lower, but again equal. There is no shift of solute.

9.4 Is it necessary to know the relative sizes of the intracellular and extra-cellular compartments in order to calculate the final osmolality following the drink? Calculate it from the initial osmolality of 285 mosmol/kg water and increase in total water from 42 to 43 litres.

The next change to consider is the addition of solute to the extracellular fluid, let us say 300 mosmol, added without water. This extra solute is assumed to remain in the extracellular fluid, as might be expected of sodium chloride (but not urea). It raises the amount of extracellular solute to 4290 mosmol (i.e. 3990 + 300) and the total amount in the body to 12 270 mosmol (i.e. 7980 + 4290). The total mass of water remains at 42 L. Now the osmolality is

$$\frac{12\,270 \text{ mosmol}}{42 \text{ L}} = 292.1 \text{ mosmol/kg water.}$$

The amount of cell water is

$$\frac{7980 \text{ mosmol}}{292.1 \text{ mosmol/kg water}} = 27.3 \text{ kg water.}$$

Thus the intracellular compartment has shrunk by 0.7 kg of water and the same amount has shifted into the extracellular compartment. The final situation is shown in Figure 9.3. The extra solute in this example could be sodium chloride, in which case the raised osmolality would be accompanied by a raised extracellular

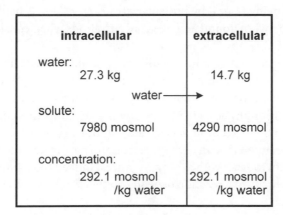

intracellular	extracellular
water:	
27.3 kg	14.7 kg
water⟶	
solute:	
7980 mosmol	4290 mosmol
concentration:	
292.1 mosmol /kg water	292.1 mosmol /kg water

Figure 9.3 In this version of the model body 300 mosmoles of solute have been added to the extracellular fluid. This draws water from the cells and raises the concentration in both compartments.

sodium concentration. It could also be, say, glucose, since this penetrates cells slowly, but then the extracellular sodium concentration would be lowered by the water that leaves the cells. On the definitions given above [Eq. (9.1) and (9.2)], the osmolar gap would be unaffected. As already noted, urea would not have the same effect, because it spreads rapidly throughout the body water; the calculations relating to Figures 9.1–9.3 are for effective osmolalities.

> **9.5**   In relation to Figure 9.3, if the extra solute did not include sodium, and there was no redistribution of sodium between body compartments, what would be the fractional reduction in extracellular sodium concentration?

Such calculations are easily and exactly carried out with model data. However, for real individuals, the sizes of the two fluid compartments are not generally known and there may be significant time-dependent complications that are not included in the models. Nevertheless the calculations do provide an approximate quantitative guide to what happens in real people.

9.10 Free and bound concentrations, activities

Most concentrations we deal with are known either through direct chemical analysis or because substances have at some stage been weighed out. These concentrations are 'total concentrations' in that a proportion of the substance in question may be bound, i.e. combined with some other substance. It is, for example, important to remember that, of the calcium inplasma (ca. 2.5 mM), only about half is 'free', the rest being bound by proteins, citrate, etc.* One is naturally aware of the distinction between bound and free concentrations when considering oxygen carriage by blood.

In the context of both chemical equilibria and physical effects, such as those of inorganic ions on membrane potentials, it is free concentrations that matter. However, there is another important concept here, namely 'activity'. Sidestepping the complications of physical chemistry, one may call this the 'effective concentration' of the solute. Except in very dilute solutions (at what is called 'infinite dilution'), solutes interact with each other in non-specific ways that generally decrease their ability to react with other substances. Indeed the Law of Mass Action states that the velocity of a reaction depends on the *activities* of the reactants. The

---

*Some physiological salines (e.g. Krebs solution) contain 2.5 mM calcium, even when there is no protein to bind any of it.

activities of inorganic ions are notably reduced in the presence of high concentrations of ions, whether these be of the same or of different species. This is because anions and cations tend to attract each other electrostatically. They even form loose ion pairs such as between sodium and carbonate or sulphate ions ($NaCO_3^-$ or $NaSO_4^-$) (see Notes). Perhaps it helps to think of a person being less effective in their activities when jostled by nearby members of the opposite sex. Strictly, however, it is not possible to define the activity of a single ionic species, as anions and cations must be considered together. In very dilute solutions activities are the same as concentrations.

Because of conceptual and practical difficulties, activities feature rather little in mainstream biomedical science. However, hydrogen ion activity is important in relation to the definition of pH (Chapter 21).

Part III

10

Graphs, straight lines and equations

There are three parts to this chapter. It begins with some elementary advice on drawing graphs. Next the general equation of a straight line is described, with some practical guidance regarding its parameters. Finally, there is a general discussion of the different types of formulae and equations one may encounter, with a warning about regression equations and how they can sometimes mislead. Logarithmic scales for graphs are discussed in Chapter 18.

10.1 Graphs: some terminology

Figure 10.1 illustrates the meanings of some words relating to graphs. The one point plotted there corresponds to $x = 0.6$ and $y = 7$. These two numbers are the 'coordinates' of the point, conventionally written in brackets as '(0.6, 7)'. The point with the coordinates '(0, 0)' is called the *origin*. As indicated in Figure 10.1, the terms 'abscissa' and 'axis of abscissae' are sometimes used for the x-axis, with 'ordinate' or 'axis of ordinates' used for the y-axis. The point where a line or curve cuts one of the axes is known as the 'intercept'. An intercept on the y-axis is the value of y when $x = 0$. An intercept on the x-axis is the value of x when $y = 0$.

Biomedical Calculations   Richard F. Burton
© 2008 John Wiley & Sons, Ltd

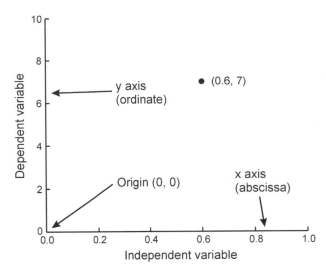

Figure 10.1   General features of a graph. One point is plotted, with coordinates (0.6, 7)

10.2 Advice on drawing graphs

This advice relates particularly to the use of graph paper, but much of it applies to computer-drawn graphs too. Before starting on the task, decide which variable corresponds to the horizontal axis ('*x*-axis') and which to the vertical axis ('*y*-axis'). Often one variable (called the 'dependent variable') is determined by, or depends on, the other (called the 'independent variable'). In that case the dependent variable is allotted the *y*-axis (Figure 10.1).

Decide whether it is necessary to include the origin, i.e. whether zero needs to be represented on both scales. Establish the appropriate range of values for each scale and mark these temporarily on the axes. Label the axes, with units. Start by plotting in pencil, so that corrections are easily made.

10.3 The equation of a straight line

A straight line is described by an equation of form:

$$y = mx + c, \tag{10.1}$$

where '*m*' is a constant corresponding to the slope or gradient, and '*c*' is another constant. The latter is the value of *y* when *x* is zero, which therefore corresponds to the intercept on the *y*-axis. This use of the letters *m* and *c* is traditional in general descriptions of the equation. If the line slopes downwards from left to right, the gradient is negative. As well as being defined by its equation, a straight line is also defined uniquely by any two points that lie on it (i.e. their coordinates).

Several kinds of task may be called for in relation to a given straight line or its equation. The following four are discussed below.

1. determining the equation of a line that passes through two specified points;

2. drawing on a graph the line that is defined by a specified equation, that is to say by particular values of m and c;

3. finding the equation of a line from its gradient and the coordinates of a single point on the line;

4. finding the line that best fits a number of points when these lie only roughly in a straight line.

10.4 Finding the equation of a line that passes through two specified points

Let us take the points as having the coordinates (x_1, y_1) and (x_2, y_2). First the gradient m is calculated either as

$$\frac{y_2 - y_1}{x_2 - x_1}$$

or as

$$\frac{y_1 - y_2}{x_1 - x_2}.$$

To calculate c this value of m is inserted into the equation along with the coordinates of either one of the points. If x_1 and y_1 are chosen, one gets

$$y_1 = (\text{calculated } m) \times x_1 + c.$$

This rearranges to

$$c = y_1 - (\text{calculated } m)x_1.$$

Since x_1, y_1 and m are known, so now is c.

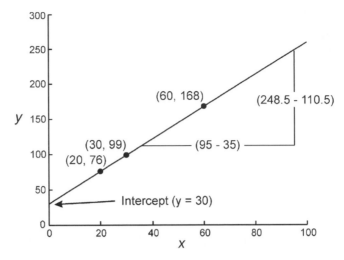

Figure 10.2 This graph shows several points lying along a straight line. Their coordinates are shown in brackets. The horizontal and vertical lines are used to illustrate the calculation of the line's gradient

By way of example, Figure 10.2 shows a line with a positive gradient that passes through a number of points for which the coordinates are shown. Two points have been selected for the calculation of the equation, namely (35, 110.5) and (95, 248.5). Horizontal and vertical lines are drawn, each through one of these points, and their lengths, corresponding to $(x_1 - x_2)$ and $(y_2 - y_1)$, are $(95 - 35) = 60$ and $(248.5 - 110.5) = 138$. Therefore

$$m = 138/60 = 2.3.$$

Calculated as above,

$$c = 110.5 - 2.3 \times 35 = 30$$

or

$$c = 248.5 - 2.3 \times 95 = 30.$$

As a check, note that when $x = 0$, $y = 30$, which inspection of Figure 10.2 shows to be true.

Check for yourself that another pair of coordinates gives the same gradient, then check that the equation works for the remaining point. It is always wise to do this if there is any chance of miscalculation.

Note that

$$\frac{y_2 \; - \; y_1}{x_2 \; - \; x_1}$$

and

$$\frac{y_1 \; - \; y_2}{x_1 \; - \; x_2}$$

may be written as $\Delta y / \Delta x$, where the symbol Δ means 'difference between values of' or 'change in values of'.

It may be that a pair of coordinates is not actually specified, but must be read off a graph that is already drawn. Then the accuracy with which the coordinates can be read may be limited. Choosing the points may be partly a matter of convenience and partly one of obtaining an accurate gradient. If one chooses a point lying on one or other axis, the arithmetic is easier in that it involves a zero. Moreover, the coordinates of points lying on one or other axis or else on a gridline (if there are ruled gridlines) may be more accurately read. If there are no gridlines, it can help to use the right-angled corner of a sheet of paper as a set-square, with one edge aligned on one axis. Then the position of one point can be marked on the other edge, prior to aligning that edge with the other axis for reading against its scale. Given the inevitable imprecision of the method, it helps to choose two points that are as far apart as convenient. This ensures that the errors in reading x_1, y_1, x_2 and y_2 have least effect on the differences $(x_1 - x_2)$ and $(y_1 - y_2)$.

10.1   Three points on a straight line have been read off a graph – but only to the nearest whole numbers. Their coordinates are (3, 3), (5, 6) and (99, 123). Calculate m from the first two pairs of coordinates and also from the first and third pairs. Comment on the difference.

10.5 Drawing a line that is defined by a specified equation

This is a matter of finding the coordinates of two points through which to draw the line. Again it helps, for accuracy, to have these points as far apart as possible. The method is to choose two values of x and then to calculate the corresponding values of y using the specified values of m and c. It may be convenient if one of the x values is zero so that y equals c and the point lies on the y-axis.

10.6 Finding the equation of a line from its gradient and the coordinates of a single point on it

Let us say, by way of example, that the gradient is 0.6 and the point is (10, 15). These numbers can be substituted into Eq. (10.1):

$$15 = 0.6 \times 10 + c = 6 + c.$$

Then, by rearrangement,

$$c = 15 - 6 = 9.$$

Therefore the equation is

$$y = 0.6x + 9.$$

If you prefer to see the procedure presented as a formula, then here are two versions to choose from, with x_1 and y_1 being the coordinates of the known point:

$$y - y_1 = m(x - x_1),$$

and

$$y = mx + (y_1 - mx_1).$$

10.7 Finding the line that best fits a number of points when these lie only roughly in a straight line

Depending on the context, it may suffice to draw a line through the points by eye and then to proceed as above. At other times the line of best fit needs to be found by a statistical technique. Of these, the one most often used is least-squares regression analysis. I leave the details of this method to statistics textbooks, but do discuss below a particular limitation of this method.

10.8 'Proportional' and 'inversely proportional'

When a straight-line relationship is such that $c = 0$, then the line goes through the origin. Then y may be described as 'being proportional to' x. Alternatively, y and x may be described as being 'proportional to' each other. The implication is that, when one variable changes, the other changes by the same factor ('in the same proportion'). The equation is now simply

$$y = mx.$$

The constant m (though in practice usually represented by some more appropriate symbol) is the 'constant of proportionality'. In other situations y may fall as x rises. Two variables are described as 'inversely proportional' to each other only when their product (xy) is constant. If the product is denoted k, the equation is

$$xy = k,$$

or

$$y = k/x.$$

10.9 Gradients of curves

Many relationships are not represented by straight lines, but by curves with continuously changing gradients. Nevertheless the gradient can be found at any particular value of x as $\Delta y/\Delta x$. In principle this requires infinitesimal values of Δy and Δx to be measured, or at least very small ones, and this can be hard to do accurately. A common procedure is to draw a straight line as a tangent to the curve at a point of interest and find the gradient of that (Figure 10.3). This is often accurate enough.

10.10 A note on units

Most of the quantities we deal with have units, but the only mention of them so far is the important one that they should be shown in the labelling of axes. The statistical calculations of regression analysis do not include them. The units for c are always the same as those for y. Those for m are always those for y divided by those for x – as is obviously so if m is calculated as a difference in y values

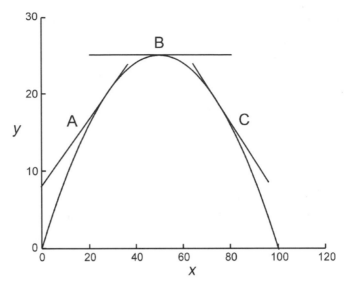

Figure 10.3 An arbitrary example of a curved relationship between y and x. The gradient, $\Delta y / \Delta x$, varies continuously, but corresponds at three points to the gradients of the straight lines that are drawn there as tangents. The gradients of lines A, B and C are respectively positive, zero and negative

divided by a difference in x values. Suppose that the distance in kilometres that a vehicle moves (y) is plotted against time in minutes (x). The gradient has the unit km/min – as is appropriate to speed.

10.11 On the different kinds of formulae and equations

Equations and formulae come in a variety of different types. Although there is little point to a formal classification, it is as well to be clear about some of the differences.* There are equations that state unarguable truths (e.g. $3 = 2 + 1$, $x^3 = x^2 \times x$). These are often a matter of definition, for example relating the different temperature scales, relating pounds to kilograms, body mass index to body mass and height. There are exact equations of physics, based on reason and experiment. There are formulae derived logically that allow one particular variable to be calculated from measurements of others, for example cardiac output

---

* Mathematicians naturally recognize a great variety of equation types that have no place in this book.

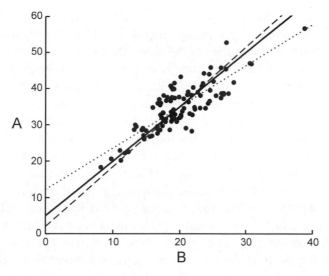

Figure 10.4 Artificial data on A and B. The thick diagonal line shows the known true relationship between them. The other two lines have been calculated by the statistical technique of least-squares regression

from measurements of oxygen consumption and oxygen concentrations in blood (Section 13.5).

There are also empirical equations. These are equations based just on observational data and without much theoretical foundation. Some of them are arbitrarily chosen equations that happen to describe the data well. Sometimes they are inexact in the sense that the data shows significant scatter, so that the equation needs to be fitted by statistical means. Expecting empirical equations of either sort to stand up to unit analysis is often pointless – and nothing to be troubled by.

Look now at Figure 10.4, but ignoring the caption for the moment. The graph shows the relationship between two unnamed variables A and B. It looks as though the relationship is linear and can therefore be described in terms of a straight line using an equation of the form:

$$A = mB + c,$$

where m and c are constants. If one applies statistics to find the best values of m and c, but knows nothing else about A and B, one has arrived at a mathematical description of an empirical relationship. As to determining the line of 'best fit', the usual method is to apply least-squares regression analysis. This is a matter for textbooks of statistics, but my objective now is to make a different point.

This is something that is not generally covered in elementary statistics courses, but which is important to a topic in Chapter 12. Often one of two correlated variables is clearly a determinant of the other; the first is 'independent' and the second is 'dependent'. Thus the diameter of a nerve fibre is one determinant of conduction velocity and not vice versa. However, sometimes variables are clearly correlated, but neither determines the other – like arm and leg lengths. Let us assume that this applies to A and B in Figure 10.4. One can calculate the regression of A on B and also separately calculate the regression of B on A and the resulting lines are both shown in Figure 10.4. The first represents the most reliable equation for estimating A when only B is known and the second represents the most reliable equation for estimating B when only A is known. They are different, so they cannot both represent the 'true' relationship between A and B, if there is one. I happen to know that there is such a true relationship and also what it is. This is so because I told my computer how to generate the numbers (see Notes). The true relationship is represented by the thick diagonal line in Figure 10.4. According to this,

$$A \approx 1.5B + 5,$$

except that there are random elements introduced that make it inexact. In other words, the equation is a precise description of an imprecise relationship. Unfortunately, without further information, *there is no statistical method for determining the 'true' equation exactly from the data* (see Notes). In some situations one of the regression lines does provide its best estimate, but that is not true in this case.

Sometimes, when there is this kind of statistical uncertainty, unit analysis can come to the rescue – or else the physicists' dimensional analysis, which amounts to much the same thing (Chapter 1). Indeed, it can happen that empirical relationships are lifted to a higher status in this way. This is illustrated in Chapter 12.

Problems

10.1   See Section 10.4.

10.2   A line has been fitted to a set of x and y values. It passes through the point corresponding to the respective mean values, which are 40 g and 150 mg. The gradient is 9.0 mg/g. (a) Calculate c. (b) At what value of x is y zero?

10.3   Variables M and N are plotted on a graph with N as the dependent variable. The origin is included in the graph. M and N range respectively between about 0 and 5 and between about 1 and 6. They are related according to

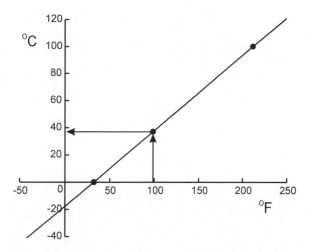

Figure 10.5   The relationship between the Celsius and Fahrenheit temperature scales. Two points indicate the freezing and boiling points of water

the following equation (a regression equation perhaps):

$$N = 0.80M + 2.0.$$

In order to add this line to the graph, two points are needed, fairly far apart, but within the ranges specified. It seems appropriate to choose values of M of, say, 0 and 4. Calculate the respective values of N.

10.4   Rearrange the equation $x = 2.4y - 3$ into the form '$y = mx + c$'.

10.5   Figure 10.5 shows the linear relationship between temperatures on the Celsius and Fahrenheit scales. Although one may use the graph to convert either way, the x-axis is allotted to Fahrenheit because scientists have more need to convert Fahrenheit to Celsius. The coordinates of two points on the line are well known, namely (32 °F, 0 °C) and (212 °F, 100 °C). (a) From those coordinates calculate the gradient and the coordinates of the point at which the line cuts the y-axis, i.e. the y-intercept. (Remember to include the units.) (b) Use these values to convert 98.6 °F (a normal core body temperature) to the Celsius scale. To check your answer, albeit only approximately, you may follow the arrows on Figure 10.5. (c) Without using the graph find the temperature at which Fahrenheit and Celsius are numerically equal. (d) What would the gradient be if the axes were reversed? (e) You probably know already the values of the two gradients, but as the ratios of two one-digit integers. What are they?

11

On shapes and sizes

This chapter includes calculations relating to simple shapes, like cylinders, spheres and cuboids, but there are also exercises on unit analysis, interpreting graphs, rearranging and checking equations, working with powers, eliminating unwanted variables and checking algebraic working. As to biological content, there is a discussion of some properties of erythrocytes, surface/volume and surface/mass relationships in general, the relationship of resting metabolic rate to body size and the calculation of the surface area of the human body.

11.1 Areas and volumes of simple shapes

Most readers will not need to be reminded that the area of a rectangle of sides A and B is AB, that the volume of cuboid with side lengths A, B and C is ABC and that its surface area is the sum of the areas of its six rectangular faces ($2AB + 2BC + 2AC$). The box below is a reminder of some other important geometrical formulae. Those all include π, which has the value $3.14159\ldots$, but for which the approximations 3.14 and 22/7 ($= 3.1429$) are adequate for most biological purposes.

> Area of circle $= \pi\,(\text{radius})^2$
> Circumference of circle $= 2\pi\,(\text{radius})$
> Volume of cylinder $= \pi\,(\text{radius})^2 \times \text{length}$
> Surface area of cylinder $= 2\pi\,(\text{radius})^2 + 2\pi\,(\text{radius}) \times \text{length}$
> Volume of sphere $= 4/3\pi\,(\text{radius})^3$
> Surface area of sphere $= 4\pi\,(\text{radius})^2$

Biomedical Calculations   Richard F. Burton
© 2008 John Wiley & Sons, Ltd

The formula shown for the area of a cylinder includes the areas of the two circles at the ends, each of area $\pi(\text{radius})^2$. As for the rest of its area, equal to $2\pi(\text{radius}) \times \text{length}$, this may be thought of as the area of a rolled-up rectangle with sides equal to the circumference and length of the cylinder.

11.1   Treating an eyeball as a perfect sphere with a diameter of 25 mm (close to 1 inch), calculate its volume.

11.2   Mitochondria are typically about 0.2–1 µm across and 2–8 µm long. Treating it as a cylinder with flat ends, calculate the volume of one that is 4 µm long and 0.4 µm across.

The formulae for ellipses and ellipsoids are less often useful, but may be noted for reference. The area of an ellipse with semi-axes a and b (equivalent to radii) is πab. The volume of an ellipsoid is $\frac{4}{3}\pi abc$, where a, b and c are the three semi-axes. The surface area of an ellipsoid cannot be expressed exactly by an elementary function, but Thomsen (2004) has given the following formula that is correct to within 1.061 per cent:

$$\text{surface area} \approx 4\pi\left[(a^p b^p + a^p c^p + b^p c^p)/3\right]^{1/p}, \qquad (11.1)$$

where $p = 1.6075$.*

11.3   Check that the formula for the surface area of an ellipsoid reduces to that for a sphere when the three semi-axes are equal.

11.2 Erythrocytes, cylinders and spheres

As an exercise relating to the volumes and surface areas of cylinders and spheres, we look here at certain properties of erythrocytes, including how they behave in solutions of reduced osmotic pressure. Of course, as a biconcave disc, an erythrocyte is normally neither of these shapes, but we can liken one simplistically to a flat cylinder (i.e. a thin disc) for the purpose of making some approximate calculations.

---

* $1.6075 = \ln(3)/\ln(2)$. Natural logarithms ('ln') are explained in Section 18.6.

Let us start by estimating a typical erythrocyte volume, taking representative text-book values for the male haematocrit (say 44 per cent after correction for plasma trapped amongst the cells after centrifugation) and for the number in 1 mm^3 (say 5.4 million).

11.4   What do these figures suggest is the average volume of an erythrocyte in μm^3 and fL?

The answer, the so-called 'mean corpuscular volume', is typically in the range 80–96 fL (where 1 fL $= 1$ μm^3).

For the purpose of our calculations, let us take it that a typical erythrocyte has a volume of about 88 μm^3 (in the middle of that range) and a diameter of 8.1 μm.[†] The area of a circle with a diameter of 8.1 μm is calculated as

$$\pi(\text{radius})^2 = \pi(8.1\mu m/2)^2 = 51.5\mu m^2.$$

11.5   Treating the erythrocyte for simplicity as a cylindrical disc with a volume of 88 μm^3 and having an even thickness (rather than being a biconcave disc), calculate its thickness.

The actual maximum thickness is greater, of course, being about 2.4 μm. The minimum thickness is about 1.0 μm.

Let us now estimate the surface area. For our flat disc this is largely made up of the combined area of the two circular faces, each of which is 51.5 μm^2, as just calculated. To this total area of 103 μm^2 must be added the area of the side of the cylinder. This equals the circumference multiplied by the height. The circumference is $2\pi \times$ radius or $\pi \times$ diameter, namely 25.4 μm.

11.6   Taking the height of this cylindrical approximation to an erythrocyte as 1.71 μm, calculate this lateral area and also the total surface area.

[†] Erythrocytes are useful indicators of scale in histological slides, but, as a result of shrinkage, they are usually nearer 7.5 μm across. Normal sizes vary and people disagree on their average dimensions.

Although the erythrocyte is actually a biconcave disc, our calculated surface area is within the range of published estimates and it is therefore used in the calculations that follow.

11.3 The swelling of erythrocytes in hypo-osmotic solutions

If erythrocytes (in heparinized blood) are suspended in a salt solution with an osmotic pressure lower than normal, they take up water and swell. At a particular critical osmolality, a given erythrocyte swells to become spherical, without change in surface area. With further swelling, the membrane is stretched and the cell contents are released, the erythrocyte then being described as haemolysed. This happens in tests of 'osmotic fragility', where erythrocytes are placed in dilute salt solutions in order to establish the mean concentration at which haemolysis occurs. The answer to the last problem may be used to calculate the size of the resulting sphere – first the radius and then, from that, the volume.

11.7   What will volume be when the erythrocyte just becomes spherical? (Start by finding the radius of the sphere.) The total surface area was calculated above as 146 μm^2.

11.8   By what factor has the volume increased?

    This increase in volume is a result of osmotic swelling, so it should be possible to relate it quantitatively to the change in osmotic pressure. In testing blood for fragility, erythrocytes are commonly suspended initially in 0.9 per cent NaCl. They are then re-suspended in more dilute solutions to find the concentration at which 50 per cent of the red cells haemolyse. With normal blood that concentration is in the region of 0.4 per cent NaCl.

    If the erythrocyte consisted simply of a cell membrane containing only inorganic ions and if there were no exchanges of solutes across its membrane, its volume up to the point of haemolysis would be inversely proportional to the external osmolality or salt concentration. This is because, following rapid osmotic equilibration, the internal and external osmolalities would remain equal and, in the absence of solute movement, this could only be accomplished by water movement and volume change. According to the last calculation, the cell swells by a factor of 1.89 in becoming spherical, so the external salt concentration would need to be reduced by a factor of $1/1.89 = 0.53$. This would correspond to a concentration of $0.53 \times 0.9\%$ NaCl $= 0.48\%$ NaCl. That is substantially above the 0.4 per cent NaCl that is more usual, but the situation described is not quite right anyway. The erythrocyte actually contains not just salts and water, but also a substantial amount of haemoglobin and other protein and some of the water is bound to these. The proteins contribute significantly to the total volume. In the normal cell the material that is not free water is about 40 per cent of the total volume and 40 per cent of our chosen total volume of 88 μm^3 is 35 μm^3. Accordingly, the volume of free water in the erythrocyte in 0.9 per cent NaCl is $(88 - 35) = 53$ μm^3 and the volume present when the erythrocyte is in the dilute solution is $(166 - 35) = 131$ μm^3. What would the concentration of that dilute solution need to be to account for this difference in volume? Let us call the answer x per cent NaCl. If these volumes are taken to be inversely proportional to concentration, then 0.9 per cent NaCl $\times$ 53 μm^3 equals x per cent NaCl $\times$ 131 μm^3.

11.9   What is x, the concentration of the dilute salt solution?

This answer is somewhat below 0.4 per cent NaCl, but there are several uncertainties and approximations in the overall sequence of calculations, so that the disparity is unsurprising. Partly because osmotic haemolysis is affected by factors such as temperature and pH, it is hard to assemble from the literature a completely consistent and satisfactory set of measurements to use in the calculations.

11.4 Distortion of erythrocytes in passing along narrow blood vessels

Erythrocytes easily distort as they pass through narrow tubes. Thus they have been seen to fold like crepe suzettes on entering a pipette tip of diameter 5 μm. There is a minimum width of vessel that an erythrocyte can traverse without its membrane stretching and this can be calculated. The shape has to be that of a hot-dog sausage, a cylinder with hemispherical ends (Figure 11.1). Both the surface area and the volume are those of the cylindrical portion plus that of two hemispheres (i.e. a cylinder and one whole sphere). The radius, R, is the same for both. Therefore, if the length of the cylindrical portion is L,

$$\text{area} = 2\pi RL + 4\pi R^2. \tag{11.2}$$

$$\text{volume} = \pi R^2 L + {}^4/_3\pi R^3. \tag{11.3}$$

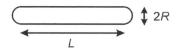

Figure 11.1 A cylinder with hemispherical ends, the supposed shape of an erythrocyte passing through a narrow capillary.

It is the radius that interests us and not the length, but the two equations can be combined, with the elimination of L, to produce the following:

$$^2/_3\pi R^3 - \text{area·}R/2 + \text{volume} = 0.$$

For the area and volume previously used, namely 146 μm^2 and 88 μm^3, this becomes

$$^2/_3\pi R^3 - 146·R/2 + 88 = 0.$$

The solution of this equation is $R = 1.25$. In other words, the minimum diameter of blood vessel through which the erythrocyte can pass is 2.5 μm. Different estimates of an erythrocyte's volume and surface area give different estimates of the minimum diameter (see Notes).

I have hurried over the mathematical details to reach this biological conclusion. Now they must be spelt out.

11.5 An exercise in rearranging equations to eliminate an unwanted term

The task is to eliminate L from the two equations given above, namely Eqs (11.2) and (11.3):

$$\text{area} = 2\pi RL + 4\pi R^2.$$
$$\text{volume} = \pi R^2 L + {}^4/_3\pi R^3.$$

For readers who lack confidence in such procedures, the steps are spelt out in detail, following the principle that whatever one does to one side of an equation should be done to the other also. To eliminate L, we need to rearrange each equation so that L is on its own on one side. Then the two resulting expressions for L may be equated.

We start with Eq. (11.2), for area. First, $4\pi R^2$ is subtracted from both sides of the equation:

$$\text{area} - 4\pi R^2 = 2\pi RL + 4\pi R^2 - 4\pi R^2 = 2\pi RL.$$

Next, the left and right sides are both divided by $2\pi R$:

$$\frac{\text{area} - 4\pi R^2}{2\pi R} = \frac{2\pi RL}{2\pi R} = L.$$

Therefore,

$$L = \frac{\text{area} - 4\pi R^2}{2\pi R} = \text{area}/(2\pi R) - 2R.$$

The volume equation has now to be treated similarly. Following the same procedure,

$$\text{volume} - {}^4/_3 \pi R^3 = \pi R^2 L.$$

$$\frac{\text{volume} - {}^4/_3 \pi R^3}{\pi R^2} = \text{volume}/\pi R^2 - {}^4/_3 R = L.$$

Since there are now two separate expressions that each equal L, they may be equated:

$$\text{volume}/\pi R^2 - {}^4/_3 R = \text{area}/2\pi R - 2R.$$

With $[\text{area}/2\pi R - 2R]$ subtracted from both sides, and both sides then multiplied by πR^2, we get, with straightforward rearrangement:

$${}^2/_3 \pi R^3 - \text{area} \cdot R/2 + \text{volume} = 0 \tag{11.4}$$

and that is the required equation. It was into this that the erythrocyte area and volume were inserted to calculate R as 2.5 μm. How this final equation may be solved to obtain R is discussed below.

11.6 Easy and general ways to check algebraic working

If one is unused to this kind of algebraic working or if, in some other example, the working is much more complicated, it would be wise to check it. Simply to repeat the same process can just result in a repetition of mistakes, but other checks are possible that can be quicker and need less thought.

1. Check the final equation by unit analysis. Here I choose to illustrate the example using metres. In the case of the equation just arrived at, the units for $^2/_3\pi R^3$ must be m^3 and that is true also for area$\cdot R/2$ and volume. This test therefore works out satisfactorily, but it is far from being a complete check.

2. Another and more conclusive method is to give some of the variables numerical values (without units) and then check the equations for consistency. In the above example, one might arbitrarily take R as 3 and L as 5. Then the area is $[2\pi RL + 4\pi R^2] = [30\pi + 36\pi] = 66\pi$. There is no need to use the numerical value of π. Similarly the volume is $[\pi R^2 L + ^4/_3\pi R^3] = [45\pi + 36\pi] = 81\pi$. Now these values (except that of L, which is not needed) are inserted into the expression on the left side of Eq. (11.4):

$$^2/_3\pi R^3 - \text{area} \cdot R/2 + \text{volume} = \ ^2/_3\pi 3^3 - 66\pi \times 3/2 + 81\pi = 0.$$

Since the right side of the equation is also zero, the check has proved satisfactory. If that had not been the case, then the same numerical check could be applied at every stage of the algebraic working until the position of the error is revealed. Any numbers could have been chosen, and not necessarily 3 and 5. However, it is best to avoid 0, 1 and 2 as these may give a false reassurance. Thus, to take a variable – call it 'x' – as being 2 does not distinguish between $(x + x)$, $(x \times x)$ and x^x. There is one general point to consider in choosing these numbers, though it is not relevant in this particular example: if a square root is involved, it can happen that the chosen combination leads to the square root of a negative number. In that case, choose again.

> Checking algebra with arbitrary numbers:
>
> It is easy,
>
>   but avoid using 0, 1 and 2!

11.7 Solving the equation by trial and error in a spreadsheet

Here again is the equation to be solved:

$$^2/_3\pi R^3 - 146{\cdot}R/2 + 88 = 0.$$

The problem is that this equation is a cubic. It may be solved by Newton's method, but many readers will be unfamiliar with that. However, there is the inelegant alternative of trial and error. This is not at all laborious if the expression on the left of the equation is set up in a spreadsheet, such as Microsoft Excel. Then one merely has to type in different values of R until the expression approximates to zero. One obviously starts by trying out a value less than 8 μm.

   The use of a spreadsheet in this way is invaluable in any context where there are many similar calculations to perform. A formula may be written out in one cell with one or more of the numerical quantities placed in separate cells for easy alteration. In this particular case the possible values of R to be tested would be entered in one cell and the formula, written out in another cell, would refer to this. In the case of Excel, the formula might be entered like this:

$$= 2/3{}^*PI(){}^*A3{}^{\wedge}3 - 146{}^*A3/2 + 88$$

where 'PI()' stands for π and 'A3' is the cell containing the value of R to be tested. Readers should become familiar with the uses of spreadsheets for calculations, but this is not the place to say more about them.

11.8 Why do we not have naturally spherical erythrocytes?

It would obviously be physically possible for the contents of a human erythrocyte to be contained in a sphere of the same volume. One disadvantage is that oxygen and carbon dioxide would have further to diffuse between the centre and the outside. Another is that only a little osmotic swelling would cause haemolysis. Indeed, any other kind of significant physical distortion could cause haemolysis too, because a sphere is the shape that minimizes the ratio of surface area to volume. The erythrocyte of the previous calculation has a volume of 88 μm^3. It is easy to calculate what its diameter would be if it were spherical and had the same volume.

11.10   For the same volume of 88 μm³, what would the diameter be if the erythrocyte were spherical and what would be its surface area?

11.11   What is the minimum diameter of vessel that this spherical ery-throcyte could pass through? The answer should be obvious from what has already been said. Nevertheless check your conclusion by inserting the rele-vant values for the erythrocyte's radius, area and volume into Eq. (11.4).

11.9 General properties of simple geometrical shapes

Because of its simplicity, we start with the upright square cuboid of Figure 11.2. Its height is H and the equal sides of its top and bottom faces are of length l. Then,

$$\text{volume} = l^2 H. \qquad (11.5)$$

The surface area is the total for the six faces of the cuboid, so that:

$$\text{surface area} = 4lH + 2l^2. \qquad (11.6)$$

Suppose now that there are many of these cuboids of different sizes, but all with the same proportions, so that the ratio of l to H has a constant value K.

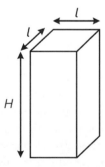

Figure 11.2   An upright square cuboid of height H and side lengths l.

Then $l = KH$. Thus:

$$\text{volume} = K^2 H^3.$$

and

$$\text{surface area} = 4KH^2 + 2K^2H^2 = (4K + 2K^2)H^2.$$

These equations can be simplified by writing k_1 for K^2 and k_2 for $(4K + 2K^2)$. Then

$$\text{volume} = k_1 \times H^3 \tag{11.7}$$

and

$$\text{surface area} = k_2 \times H^2. \tag{11.8}$$

Both volume and surface area rise steeply with H, as is shown graphically in Figure 11.3. The lines are calculated for a cuboid of specific proportions and so for specific values of K. The actual values are immaterial here, but a way of finding them is illustrated below. The forms of the curves are independent of those numbers.

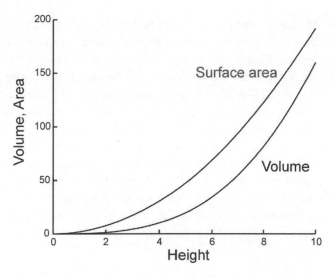

Figure 11.3   The dependence of volume and surface area on height for square cuboids of particular proportions. The units are unspecified.

These relationships may be generalized to all other shapes where the sizes vary and the proportions are constant. There is no need to consider height as such, because any convenient characteristic linear measurement will do, such as width or perimeter length etc. With this denoted L,

$$\text{volume} = k_v L^3 \qquad\qquad (11.9)$$

and

$$\text{surface area} = k_a L^2, \qquad\qquad (11.10)$$

where k_v and k_a are constants specific to the shape in question.

> For objects of identical shape, with some particular linear dimension denoted L:
>
> > volume increases with L^3,
> >
> > surface area increases with L^2
>
> and
>
> > surface/volume ratio is proportional to $1/L$.

An important and general biological point arises from these relationships. With increasing size, the surface area of any object of constant proportions increases less steeply than the volume. Put another way, the surface/volume ratio falls with increasing size. For hypothetical cells and simple organisms of constant form and proportions, the ability to take up oxygen increases with surface area, while the oxygen consumption, one may postulate for the moment, increases with the volume (i.e. the amount of tissue). There is then a disadvantage to large size that may need to be offset by anatomical or physiological specializations aiding oxygen uptake. Similar considerations apply to the exchange of other materials. A large snowman may outlive a small one because the surface area for heat exchange in the former is lower relative to the mass of snow. Relevant surfaces can include those of internal organs as well as those visible from outside. Such considerations suggest one should combine Eqs (11.9) and (11.10) to produce the following:

$$\frac{\text{surface area}}{\text{volume}} = \frac{k_a}{k_v} \cdot \frac{1}{L}. \qquad\qquad (11.11)$$

In words, the ratio of surface area to volume is inversely proportional to L. We return to this simple conclusion later.

11.10 Replacing volumes with masses in these equations

The discussion so far has involved volumes rather than masses, but it is mass that is more often measured. When density is constant, the two are exactly proportional to each other and mass is therefore often accepted as a replacement for volume, with, say, 1 g substituting for 1 cm^3 and 1 kg substituting for 1 L. With unit analysis in mind, we note, however, that the units differ. Let us therefore be more rigorous about this. To replace volumes with masses, recall that density equals mass per unit volume (i.e. mass/volume). Therefore,

$$\text{volume} = \frac{\text{mass}}{\text{density}}. \tag{11.12}$$

11.12 Is this right? Check this formula by spelling out the units in terms of metres and kilograms.

> For objects of identical shape and density, mass is proportional to volume.

This 'trick' of replacing volume with mass is often used without comment, but it can reveal itself when the unit analysis of a formula reveals that a density term is missing.

11.11 A digression on graphs

We may now take the opportunity of looking at some more aspects of the graphing of data. Suppose that one has data on volume and L for some irregular, but constant, shape and wishes to find the value of k_v graphically. A good way of achieving this is to find a way of plotting the data that produces a straight line. The two most obvious ways of doing that are to plot volume against L^3 or the cube root of volume against L. (Another method is to plot the data as their logarithms, as discussed in Chapter 18.) Figure 11.4 shows the outcome of the

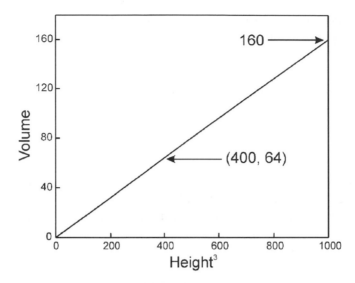

Figure 11.4 The volume of the square cuboid of Figure 11.3 plotted against height³. Two points are indicated for use in Problem 11.13.

first method as applied to the volume relationship of Figure 11.3. L is taken as the height H.

11.13   Some points on the line in Figure 11.4 are specified. Calculate the gradient. This equals k_1 in Eq. (11.7) and is therefore K^2. So what is K for the cuboid of Figures 11.3 and 11.4?

11.12 Calculating surface area from volume and height: another exercise in re-arranging equations and eliminating unwanted terms

Let us return to our square cuboid of Figure 11.1. Suppose that we know the volume, V, and also H, but not l, and that we wish to calculate the surface area, A. What is the necessary formula? (There is a point to this in regard to the human body, because body volume is approximately proportional to easily measured mass and because we deal later with the estimation of surface area from mass and height.)

As already noted,

$$\text{volume} = V = l^2 H \qquad (11.5)$$

and

$$\text{surface area} = A = 4lH + 2l^2. \qquad (11.6)$$

However, the desired formula does not include l, so it now needs to be eliminated. Taking Eq. (11.5) and dividing both sides by H, we get:

$$V/H = l^2 H/H = l^2.$$

We then take the square root of both sides:

$$\sqrt{V/H} = l.$$

Now l can be replaced in Eq. (11.6) with $\sqrt{V/H}$:

$$\begin{aligned}
A &= 4\sqrt{V/H} \cdot H + 2(\sqrt{V/H})^2 \\
&= 4\sqrt{V/H} \cdot \sqrt{H^2} + 2V/H \\
&= 4\sqrt{H^2 V/H} + 2V/H.
\end{aligned}$$

Therefore

$$A = 4\sqrt{VH} + 2V/H. \qquad (11.13)$$

This is the required equation, with l eliminated. The units for A, V and H must of course be consistent (e.g. m², m³ and m, respectively). Looking back over the derivation of this equation we see that the last term, $2V/H$, corresponds to $2l^2$ in Eq. (11.6) and to the total area of the top and bottom surfaces of the cuboid.

11.13 Another digression to check algebraic working

The above algebraic working may be checked as suggested under Section 11.6, first by unit analysis and then by the numerical method.

1. In the case of Eq. (11.13), working with metres, the units of $\sqrt{VH}$ are $(m^3{\cdot}m)^{1/2} = (m^4)^{1/2} = m^2$. The units for V/H are $m^3/m = m^2$. Both are appropriate for components of the area A.

2. The more conclusive method is that of allotting numerical values to variables and thus checking the equations for consistency. In the above example, one might arbitrarily take l as 5 and H as 7. Then, from Eq. (11.5),

$$V = 5^2 \times 7 = 175.$$

Likewise, from Eq. (11.6),

$$A = 4 \times 5 \times 7 + 2 \times 5^2 = 190.$$

Then, from Eq. (11.13),

$$A = 4 \times \sqrt{175 \times 7} + 2 \times 175/7 = 140 + 50 = 190.$$

This is the same answer for A as obtained three lines above, so the check has proved satisfactory.

11.14 Generalizing the formula to include the human body

A formula like Eq. (11.13) applies to vertical cylinders and some other geometrical shapes and can be written in the following general form:

$$A = \alpha\sqrt{VH} + \beta V/H, \qquad (11.14)$$

where α and β depend on the shape in question and have no units.

11.14   Find the values of α and β for a vertical cylinder.

Now let us consider the manikin of Figure 11.5 which is made up of six square cuboids. Equation (11.14) fits this too, but we will not go through its derivation as it is essentially the same as for the cuboid and cylinder. Excluding the arms, the total area of the horizontal surfaces (shoulders, soles of feet, etc.) equals the last

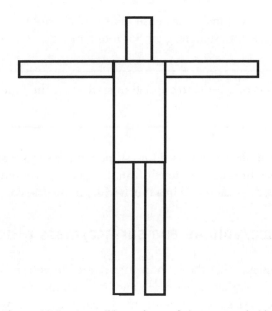

Figure 11.5 A manikin made up of six square cuboids.

term of Eq. (11.14), namely $\beta V/H$. Cylinders may be substituted for cuboids and complexities may be added for the sake of realism, with Eq. (11.14) still applying. In fact it seems likely that the formula could be applied to real human bodies to an accuracy quite acceptable in the context both of measurement accuracy and usefulness (see Notes). Appropriate values of α and β for the human body are probably about 5 and 3, respectively.

The masses of bodies are more often and more easily measured than their volumes. Therefore Eq. (11.14) may usefully be re-expressed with mass replacing volume. With ρ standing for density and W for mass, so that $V = W/\rho$, the equation becomes:

$$A = \alpha\sqrt{WH/\rho} + \beta(W/\rho)/H$$
$$= (\alpha/\sqrt{\rho})\sqrt{WH} + (\beta/\rho)W/H. \qquad (11.15)$$

For practical purposes, it is satisfactory to simplify Eq. (11.15) as

$$A = \alpha'\sqrt{WH} + \beta'W/H \qquad (11.16)$$

with density incorporated into the formerly unit-free constants α and β. Obviously α' and β' do therefore have units, but will be numerically close to α and β if the units are chosen so as to make the density numerically close to 1. That is true if the density units are, for example, g/cm^3 or tonne/m^3.

> **11.15**   In terms of centimetres and grams, what are then the units for α' and β'?

The type of formula actually used for estimating the surface area of the body from mass and height is quite different from Eq. (11.16). We return to this later after looking at body surface area in a much wider biological context.

11.15 Surface/volume and surface/mass ratios

We have noted already that the ratio of surface area to volume is important in contexts such as the exchange of oxygen and heat and that this ratio, for objects of identical shape, varies inversely with any chosen linear dimension. (The latter could be length, width, circumference, etc., because these are all proportional to each other if the shape is constant.) The surface/volume ratio is therefore central to considerations of body and cell sizes and of the need for special adaptations for exchanges with the environment. Such ideas may equally be treated in terms of mass rather than volume, and it is the surface/mass ratio that is now discussed.

Because mass (for a given shape and density) is proportional to the cube of any linear dimension and areas to its square, surface/mass ratios are inversely proportional to mass$^{1/3}$.

Let us look at the implications of this for heat loss in mammals. In order to draw a clear and useful conclusion, we may postulate two hypothetical mammals of different species that, for the purpose of calculation, have certain characteristics, not all of which may be realistic.

1. The two mammals differ 1000-fold in mass (in that respect roughly like a capybara and a suitably chosen mouse).

2. They have the same metabolic rate when that is expressed per unit body mass. This means that the larger mammal produces, and therefore loses, 1000 times more heat in a given time than does the smaller one.

3. They have exactly the same shape in all relevant details.

4. Heat is lost only through their surfaces (and not in the breath, therefore).

5. The two mammals have similar insulation, so that the rate of heat loss for a given area of surface is the same for a given temperature difference between body and air.

6. Their internal body temperatures may differ, however.

7. The rate of heat loss is proportional both to the temperature difference between body and air and to the surface area.

Since the larger mammal produces and loses heat at a rate 1000 times that of the smaller one and has only 100 times the surface area through which to lose it, there must be a bigger difference in temperature between its interior and the environment.

> **11.16** By what factor must the temperature difference exceed that in the larger mammal? If the inner temperature of the smaller one is 37 °C and the environmental temperature is 17 °C, what is the temperature inside the larger mammal?

This is such an absurdly high temperature that at least one of the original postulates must be very wrong. Given the generally greater physical activity of small mammals (scurrying mice), as compared with large ones (plodding cows), it is easy to believe that it might be the second postulate that is notably incorrect, namely the constancy of metabolic rate per unit mass. (But note that visible physical activity is hardly a good guide to resting metabolic rate!) In 1883 Max Rubner proposed a 'surface rule' according to which resting metabolic rates in different species of mammal vary in proportion to surface area, or $mass^{2/3}$. This means that the resting metabolic rate per kilogram of body mass is less in larger mammals, varying with $mass^{2/3}/mass = mass^{-1/3}$.

Much has been written about the relationship between resting (basal) metabolic rate and body mass in mammals, with the evidence suggesting over many years that resting metabolic rate is proportional to $mass^{3/4}$. This relationship is known as Kleiber's rule. However, convincing evidence has recently been put forward that the '2/3' power relationship is actually more appropriate (see Notes) and that it is in accordance with Rubner's surface rule. Crude calculations like the one above cannot give us the exact exponent, of course, for mammals vary in shape, insulation and way of life. What is clear is that basal metabolic rate in different mammals is certainly not proportional to mass.

What makes this relevant in a medical context is the desirability of matching the dosages of some drugs to a person's metabolic rate. Because this is not instantly and easily measurable, the surface area of the body is used as a guide instead. From what has been said so far, the obvious way of doing this would involve taking the area as proportional to mass to the two-thirds power. However, a more sophisticated way of estimating body surface area is needed, such as might involve Eq. (11.16). However, be warned that the actual equation that has long been used is quite different (see below). That is something that will need explanation.

11.16 The surface area of the human body

Estimates of the body's surface area are used in various clinical and physiological situations as a measure of body size. There are accepted formulae relating the area to body mass and height and these can provide useful practice in working with non-integer exponents. The formulae universally used are very different from Eq. (11.16) and their theoretical rationale has long been missing. The link between the two kinds of equation is beyond the scope of this book (see Notes).

Why estimate the surface area?

It is often necessary to consider some aspect of the size of the body. Whether a particular measured glomerular filtration rate or cardiac output is to be considered normal depends on the size of the person. To judge whether a heart is enlarged one must know how big the rest of the body is. An ideal drug dose depends both on the volume of patient into which it will be dispersed and on the typically size-dependent rate at which it will be metabolized or excreted. Whole-body metabolic rate is often a key factor here and, in mammals generally as just discussed, metabolic rate per unit mass shows a strong tendency to decrease with increasing body size (Section 11.15). Thus the resting metabolic rate of an elephant is much less per kilogram than that of a typical human and this explains the tragic death of an elephant that was injected with the drug LSD. The dose was scaled up on the basis of mass, rather than metabolic rate, from that needed to put a cat into a rage: the elephant immediately ran around trumpeting, then collapsed in convulsions five minutes later and died.

Because of the old idea that resting metabolic rate increases more or less in proportion to body surface area (i.e. Rubner's surface rule), a quick estimation of surface area is seen as clinically important. Basal metabolic rates of individuals are often expressed, for comparative purposes, in terms of surface area, e.g. as calories per square metre per hour. One may question whether human metabolic rates actually scale with surface area, but at least that is more nearly correct than that it scales with body mass. One may also question whether any physiological variable really relates to body surface area, since there are no obvious mechanisms to link them exactly. Heat loss through the skin does relate to metabolic heat production, but heat loss is uneven over the surface, being affected by regional variations in blood flow, insulation, convection and evaporation. What must be accepted is that formulae for estimating surface area from body mass and height are commonly used for indexing cardiac output, etc., and for calculating drug dosages. Inasmuch as a standardized method works, it is appropriate to use it until research establishes a better one. Scepticism aside, some readers of this book will therefore need to apply surface area formulae. There are also general lessons to be learned from them.

11.17 Standard formulae for body surface area

The earliest and best-established formula was proposed by Du Bois and Du Bois in 1916. With surface area (A) expressed in square centimetres, body mass (W) in kg and height (H) in centimetres, the formula is as follows.

$$A = 71.84 \times W^{0.425} \times H^{0.725}. \tag{11.17}$$

As an example, let us apply this formula to a 75 kg individual of height 170 cm:

$$A = 71.84 \times 75^{0.425} \times 170^{0.725}$$
$$= 71.84 \times 6.265 \times 41.41$$
$$= 18\,638\text{cm}^2.$$

Note in passing that the precision of this answer, with its five significant figures, far exceeds the exactness of any direct measurements of body surface area – and that applies also to the '71.84' in Eq. (11.17). In fact it is more usual to give the area in square metres to three significant figures, namely 1.86 m^2 in this case. (A square metre has sides of 100 cm, so that 1 m^2 = 10^4 cm^2.) As a check on the arithmetic, you may (indeed should) ask yourself if the answer seems reasonable. Thus you might consider whether 1.86 m^2 of cloth would be roughly the right amount to enclose someone completely in a tight-fitting garment. (Many of the early measurements in the nineteenth and twentieth centuries were made by coating bodies, not with cloth, but with paper.) Clinical measurements, of creatinine clearance for example, are often expressed in terms of a standard surface area of 1.73 m^2.

This may be all you need to know about the formula if you merely need to apply it and to calculate surface areas for practical use. The rest of the chapter is about applying unit analysis, obtaining practice with exponents, looking more closely at the formula and relating it to others of similar form.

You may have noticed that I present the above calculation without the units, thus flouting the rule that I have so much emphasized. So here is the calculation again, with units included:

$$A = 71.84 \times (75 \text{ kg})^{0.425} \times (170 \text{ cm})^{0.725}$$

$$= 71.84 \times 6.265 \text{ kg}^{0.425} \times 41.41 \text{ cm}^{0.725}$$

$$= 18\,638 \text{ kg}^{0.425} \text{ cm}^{0.725}.$$

That turns out to be unhelpful in that the units are not those for area (cm^2), but something rather complicated. This implies that the '71.84' is not free of units – a point seldom mentioned. We deal with this in the next two paragraphs.

There are other formulae in use that have the same general form. This may be expressed by replacing the specific numbers in Eq. (11.17) with symbols:

$$A = K \times W^a \times H^b \tag{11.18}$$

In order to analyse Eq. (11.18) in terms of units, we may consider what the appropriate units for K must be. The answer is cm$^{(2-b)}$/kg^a. That is awkward and not obviously meaningful. Perhaps that is one reason why the units for K are rarely mentioned. Trying an easier tack, let us reason as Du Bois and Du Bois did, assuming a constant body density, so that mass, W, is proportional to volume, V. More exactly,

$$W = V \times \text{density.}$$

Then

$$A = K \times \text{density}^a \times V^a \times H^b \tag{11.19}$$

Let us assume now that K has appropriate units to make the term $K \cdot \text{density}^a$ a unit-free number. Then the expression $V^a \cdot H^b$ would need to have units of area (cm^2). This can only be true if a and b are related in a particular way. Spelling out the units for this expression we have

$$(cm^3)^a \times cm^b = cm^{3a} \times cm^b = cm^{3a+b}.$$

Therefore cm^{3a+b} must be the same as cm^2, so that

$$3a + b = 2. \tag{11.20}$$

Checking Eq. (11.20) against the exponents in Eq. (11.17), we find that $(3 \times 0.425 + 0.725)$ does indeed equal 2. This is what Du Bois and Du Bois intended.

Equation (11.20) is important for the propriety of Eqs (11.17) and (11.18) in terms of unit analysis (or their dimensional correctness, one may also say), but there is another important implication. Looking at the latter two equations one might suppose that three separate constants are needed to define how surface area depends on mass and height [as compared with two with Eq. (11.16)]. However, if we combine Eqs (11.18) and (11.20), we get a formula with only two constants, namely K and a:

$$A = K \times W^a \times H^{(2-3a)} \tag{11.21}$$

11.17   There is another version in which the only two constants are K and b. What is that?

We may now return to the matter of the units for K. These are given above as $cm^{(2-b)}/kg^a$, but we can now make use of Eq. (11.20) to simplify that. Rearranging Eq. (11.20) we see that $b = 2 - 3a$. This allows b to be eliminated from $cm^{(2-b)}/kg^a$, which thus becomes cm^{3a}/kg^a, or $(kg/cm^3)^{-a}$. K therefore has units appropriate to density^{-a}. We come to the same conclusion looking at Eq. (11.19), because, as assumed above, $K \cdot density^a$ has to be free of units.

Du Bois and Du Bois based their formula on data for nine individuals, though they tested it on more. Subsequent authors have proposed different formulae that are based on other data (see Notes). Particularly favoured, partly for their simplicity, seem to be versions in which the surface area is taken as proportional to $(WH)^{0.5}$ [which becomes more suggestive of Eq. (11.18) when written as $W^{0.5} \times H^{0.5}$]. Du Bois and Du Bois tried this version themselves and their own data support it only marginally less well than they support Eq. (11.17). With measurements in kg, cm and cm^2, the best constant in the alternative formula is given as 167. This is almost exactly 1000/6, so the formula can be written in easily remembered form as:

$$A = 1000/6 \times (WH)^{0.5} \qquad (11.22)$$

11.18   Does $(3a + b)$ equal 2 in the case of Eq. (11.22)?

11.18 An exercise in comparing formulae containing exponents

In Eqs (11.17) and (11.22) we have two formulae for calculating the surface area of the body from mass and height. They cannot both be exactly correct, so how do they compare in practice? We can look at the question as an exercise in manipulating equations with exponents.

The obvious approach is to compare surface areas calculated with both equations for particular combinations of mass and height.

11.19   Using both Eqs (11.17) and (11.22), calculate the surface area, A, for the following two combinations of mass and height: (a) $W = 66$ kg, $H = 170$ cm. (b) $W = 200$ kg, $H = 170$ cm.

One could repeat that process for various combinations of mass and height, but there is another, more systematic, approach. An equation may be derived that describes all those combinations of mass and height that give identical estimates of surface area from the two equations. Combining Eqs (11.22) and (11.17), we have:

$$A = 1000/6 \times W^{0.5} \times H^{0.5} = 71.84 \times W^{0.425} \times H^{0.725}.$$

This can be seen as a statement that the two equations give the same estimates of A, but we can now mentally delete the first part '$A =$' and see the rest as defining the required relationship between W and H. That now needs to be rearranged and simplified.

Let us bring both W terms to one side and both H terms to the other side. To this end, we divide both sides by $W^{0.425}$, $H^{0.5}$ and 1000/6:

$$\frac{1000/6}{1000/6} \times \frac{W^{0.5}}{W^{0.425}} \times \frac{H^{0.5}}{H^{0.5}} = \frac{71.84}{1000/6} \times \frac{W^{0.425}}{W^{0.425}} \times \frac{H^{0.725}}{H^{0.5}}.$$

The first, third and fifth ratios each equal 1 and can be cancelled out. The fourth ratio simplifies to 0.431. Therefore,

$$\frac{W^{0.5}}{W^{0.425}} = 0.431 \times \frac{H^{0.725}}{H^{0.5}}.$$

11.20 Simplify this to show the dependence of W on H.

11.21 As a check, calculate W for any value of H (say 160 cm) using the equation just derived. Then calculate A using Eqs (11.17) and (11.22) to see if they give the same answer.

Figure 11.6 shows the relationship between mass and height, as just established, that gives identical estimates of surface area using the two equations. Figure 11.6 also shows analogous relationships between mass and height for which Eq. (11.22) gives estimates that are 5 per cent higher and 5 per cent lower than those obtained

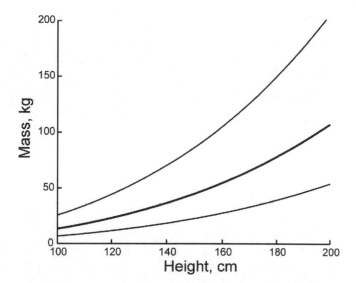

Figure 11.6 Equations (11.17) and (11.22) give identical estimates of surface area when body mass and height lie on the central curve. Equation (11.22) gives estimates 5% higher than Eq. (11.17) when mass and height lie on the upper curve and 5% lower when mass and height lie on the lower curve.

with Eq. (11.17). The equations for those two curves may be obtained in a similar manner to that already illustrated.

11.22   It has already been shown that the two equation gives different estimates of surface area for $W = 200$ Kg and $H = 170$ cm. Confirm this by finding the corresponding point in Figure 11.6.

Problems

11.1   See Section 11.1.

11.2   See Section 11.1.

11.3   See Section 11.1.

11.4   See Section 11.2.

11.5   See Section 11.2.

11.6   See Section 11.2.

11.7   See Section 11.3.

11.8   See Section 11.3.

11.9   See Section 11.3.

11.10   See Section 11.8.

11.11   See Section 11.8.

11.12   See Section 11.10.

11.13   See Section 11.11.

11.14   See Section 11.14.

11.15   See Section 11.14.

11.16   See Section 11.15.

11.17   See Section 11.17.

11.18   See Section 11.17.

11.19   See Section 11.18.

11.20   See Section 11.18.

11.21   See Section 11.18.

11.22   See Section 11.18.

11.23   A syringe barrel contains 1 mL of fluid occupying a length of 60 mm. What is the internal diameter?

11.24   What is the mean corpuscular volume for a blood sample with a haematocrit of 40 per cent (corrected for trapped plasma) and an erythrocyte count of 4.6 million per mm^3?

11.25   In relation to erythrocyte fragility a solution is specified as '0.9 per cent NaCl'. What are the sodium and the chloride concentrations in mmol/L solution? (The molar mass of NaCl is 58.5.) What is the osmosity?

11.26   This relates to an important series of experiments carried out by Gorter and Grendel in 1925 using blood of various species. In the case of the rabbit, 10 mL of blood contained erythrocytes at 6.6×10^6 per mm^3. The surface area of a red cell was estimated at 74.4 μm^2. (a) What was the total erythrocyte surface area, in m^2, in the 10 mL? (b) The membrane lipid was extracted in acetone, and a fraction of it, now in benzene, was spread as a unimolecular film on the surface of water. Then its area was reduced by lateral pressure until the lipid molecules lay snugly side by side. The area of the resulting film, scaled up to the original 10 mL of blood, was 9.8 m^2. What does a comparison of the two areas suggest?

11.27   Check on the above derivation of Eq. (11.13) using a different pair of numerical values for l and H. Proceed just as under Section 11.13.

11.28   Derive Eq. (11.13) for an upright cylinder of height H and radius r.

11.29   Size is important to the ease with which oxygen diffuses to the centre of a respiring cell, or to the interior of any organism lacking a circulatory system or other means of oxygen transport. Other factors matter too, notably shape, so compare, for simplicity, two hypothetical living 'spheres' that differ only in size. One has a radius of 10 μm, the other a radius of 10 mm. (a) By what factor do they differ in relative surface area (i.e. surface area divided by volume of respiring substance)? (b) By what factor do they differ in the diffusion distances from surface to centre?

11.30   Consider a hypothetical population of individuals varying in size. All are identical in shape and density so that W is proportional to H^3 and $H = cW^{1/3}$, where c is constant. Assume that their surface areas accord with Eq. (11.16). Eliminating H from this equation, test whether A works out as proportional to $W^{2/3}$ as it should do.

11.31   According to Eq. (11.17), what is the surface area of an 80 kg individual of height 175 cm?

11.32   According to Eq. (11.22), what is the surface area of the same 80 kg individual of height 175 cm? The mass and height are thus as in problem 11.31: does Eq. (11.22) give the same answer as Eq. (11.17)?

11.33   One may wish to use Eq. (11.17), but to express the surface area, A, in m^2 instead of in cm^2 and the height H in m instead of in cm. What does the constant 71.84 become? You can check your answer by repeating the calculation given above for a 75 kg person of height 1.70 m. The answer then was 1.864 m^2.

11.34   Consider the following equation, in which a, b and c are constants:

$$ax^2 + bx + c = 0.$$

This has the general form of a quadratic equation and typically has two solutions (roots) that may be calculated as

$$\frac{-b + \sqrt{b^2 - 4ac}}{2a} \quad \text{and} \quad \frac{-b - \sqrt{b^2 - 4ac}}{2a}.$$

The task here is not to show how these two expressions may be derived from the original equation, but to check each of the two expressions against it by replacing x, a and b with convenient numbers. Why is it best in this case to choose x, a and b and not some different combination of three symbols?

12

Body size, body build, fatness and muscularity: unit analysis as an aid to discovery

There are two main themes to this chapter. In one sense it is about how gross aspects of body form and composition vary with body size, the variables being mostly measurements that can be made on the intact human body – its height, mass and contents of fat and muscle, the thicknesses of skinfolds with their underlying adipose tissue and the ability to lift weights. The other theme is the help that unit analysis can give, not now in getting calculations right, but in thinking about relationships and formulae. There are good reasons for concentrating on the particular field of 'anthropometry' for this purpose: the variables are easy to visualize and the units – metres, kilograms, etc. – are familiar and already treated in earlier chapters. Moreover, as will be illustrated, attention to unit consistency and compatibility can lead to new insights in this field.

Related topics have been covered already, namely the implications of treating fat content as a percentage (Section 3.1), the relationship between fat content and body density (Section 4.3) and the estimation of body surface areas (Sections 11.16–11.18). Here we look at relationships between fat-free body mass and height, at a particular measure of body build – the body mass index – and then at the cross-sectional areas of skeletal muscle in relation both to body height and to weightlifting records. Finally we consider how estimates of body fat from skinfold thickness measurements might be improved. Other relevant topics

---

Biomedical Calculations   Richard F. Burton
© 2008 John Wiley & Sons, Ltd

covered earlier include area–mass relationships (Chapter 11.) and lines of 'best fit' (Section 10.11).

As said already, a particular preoccupation in this chapter is with unit compatibility and consistency in formulae. In the formulae of physics, like that for the period of a pendulum discussed in Chapter 1, the units for the expressions either side of the equals sign must be consistent – unless both expressions are unit-free, of course. This condition is often necessary or desirable in biological formulae too and, when it does not evidently pertain, it is sometimes profitable to think whether the formula in question can be improved.

Here we are much concerned with variations associated with body size. Relationships, for example between mass and height, are best revealed when the data cover wide size ranges. However, it is important to bear in mind that variations in human body size arise in two distinct ways. Thus they include those due to growth and those due, most clearly in adults, to individual variation. These must be distinguished, because growth during childhood is not just a matter of increasing size, but of changing body shape.

12.1 Variations in fat-free mass with height and age

We have seen for objects of identical shape and density that mass is proportional to the cube of any characteristic linear measurement. Given that human bodies are generally rather similar in form (and disregarding shape changes due to posture and limb movements), we may ask whether body mass actually does tend to vary with the cube of height. We come to that question later, but focus for the moment on the fat-free mass, namely the mass of the body less the mass of fat. Variations in fatness greatly increase the scatter in the relationship between total mass and height and a tighter relationship is to be expected if one considers just fat-free mass. The latter is sometimes called the lean body mass, although that term is sometimes taken to exclude bone mineral as well as fat. (Actually, fat-free mass is not completely independent of fat content because losses and gains of fat are accompanied by losses and gains of other tissue.)

As already noted, one should distinguish between adults and children. In children mass and height obviously tend to increase together, so that a relationship exists between them even for one growing individual. In adults the variations in height are mainly from individual to individual and independent of age – though height and mass do tend to decrease in the elderly. Figure 12.1 shows data for fat-free mass and height in females aged 6–86 years, with each point representing an average. At first glance, fat-free mass does seem to rise roughly in proportion to height cubed. However, there is a definite curvature, and there are significant departures from the diagonal line drawn through the points. What this means

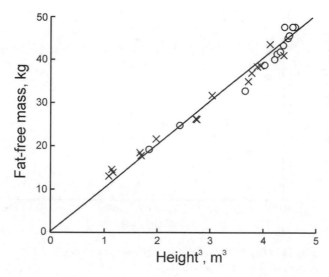

Figure 12.1   The relationship between fat-free mass and body height cubed in females. Each point represents an average for 10–129 individuals: ○ data of Barlett *et al.* (1991); × data of Ellis *et al.* (1999). The line through the origin is positioned by eye.

is that there are progressive changes in body build with increasing height (and, in children, also with age). The original data are unaccompanied by information on these, but, on the basis of changes visible and familiar in growing children, one can attempt some interpretation of the curve. Thus the fact that the points for the youngest individuals lie above the line may be attributed at least partly to relatively short leg lengths. (One way of expressing the latter is in terms of 'relative sitting height', which is the distance measured from the seat of a chair to the top of the head, divided by the total body height. This can be about two-thirds in the new-born, falling to roughly half in adults.)

A relationship like that of Figure 12.1 applies in males too, except that fat-free masses tend to be greater in adults for a given height. That probably reflects increased muscle mass.

This ratio of fat-free mass to height cubed can be taken as an index of fat-free body build. Although it does reflect more than one feature of fat-free build, it is as good an index as is available from just these two body measurements. It is true that other indices could be defined as that one's reciprocal or cube root, but those contain the same information. Figure 12.2 shows, for females, how the index (fat-free mass)/height^3 changes progressively with age. The fall in the early years is apparent, and also the tendency for fat-free mass to fall in women after about the time of menopause. (Similar trends are seen in males, but the adult decline in fat-free mass tends to start earlier.)

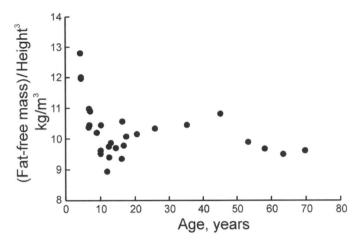

Figure 12.2   Changes in the index (fat-free mass)/height^3 with age in females. The data are from the same sources as those of Figure 12.1.

The unit used for the index in Figure 12.2 is kg/m^3. It is appropriate to cube height in order to satisfy the needs of unit analysis. However, one may still wonder what 'height^{3}' can mean in causal terms. It would make more obvious sense to think of the fat-free body volume as being proportional to height multiplied by two horizontal measurements – like the volume of an ordinary cardboard box. However, given that bodies are generally fairly similar in shape, these two unspecified horizontal measurements should be roughly proportional to height. The product of all three should therefore be roughly proportional to height^3. If height is the only linear measurement that is recorded, then it is 'height^{3}' that must be used.

Because the unit kg/m^3 is also the unit of density, dividing by density gives an index free of units and therefore independent of scale. That is appropriate for any measure just of shape. However, the inclusion of density in the index would make little practical difference, because the density of the fat-free body varies even less than that of the whole body. (Remember that fat content has a major influence on body density, as discussed in Section 4.3.)

If the average value of the index is known for a particular age group and sex, it is possible to obtain a rough estimate of the likely fat-free mass of an individual just from height. However, for statistical reasons, that is not usually done. Perhaps that is why '(fat-free mass)/height^{3}' appears to have no name.

Because of the curvature in Figure 12.1, you may be wondering why I do not fit a curve to the data instead of a straight line. Its equation, optimized by the application of statistics, would clearly have some value in predicting fat-free mass from height. It would, however, obscure the role of changing body shape in the relationship.

12.2 The Rohrer index, or 'height–weight index of build'

We have seen how the relationship between fat-free mass and height can be characterized by the index (fat-free mass)/height^3. For total body mass there is a similar index that is longer established and more often used, namely

$$\frac{\text{body mass}}{\text{height}^3}.$$

This is sometimes called the Rohrer index, or Rohrer's index. It has also been called more descriptively the 'height–weight index of build'. It is necessarily constant for geometrically similar individuals, as was noted by Quetelet as long ago as 1835.

Some of the things said earlier of the index (fat-free mass)/height^3 are also true of the Rohrer index. It too has the same unit as density, e.g. kg/m^3, so that it too becomes unit-free when divided by density (i.e. the density of the whole body this time). It too falls during childhood (Figure 12.3).

The Rohrer index differs in two important ways from its fat-free equivalent. Firstly, it is based only on measurements that are easily and accurately made, whereas fat and fat-free mass are hard to measure directly. Secondly, it includes

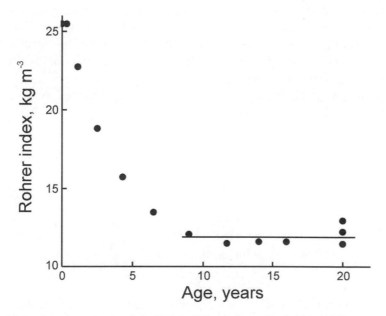

Figure 12.3 Age-dependence of the Rohrer index in boys. The horizontal line corresponds to an index of 11.9 kg/m^3. The data are from Bardeen (1920).

body fat. One might therefore suppose that the Rohrer index could be useful in predicting fatness, especially if other relevant factors such as age are taken into account. However, it is the related 'body mass index', or BMI, that is usually used for this purpose, this being calculated as (body mass)/height^2. On grounds of unit analysis, the height exponent of two may seem strange, but, as explained below, there is a good reason for it. Put simply, the BMI, with height squared, is found to be a better predictor of fatness in adults. Nevertheless, it is important to realise that the Rohrer index is a valid index in its own way. I hope that that has already been clearly shown, but long familiarity with the BMI can make that a hard point to accept.

Given data just on height and mass, the Rohrer index is the only available index of shape apart from such variants as its cube root and the reciprocal of its cube root (both of which, confusingly, have been called the 'ponderal index'). There is an important and fundamental distinction between the Rohrer index and the BMI. The BMI is a concept based in the statistics of populations and on relationships between mass and height in particular populations. The Rohrer index is a measure of body build that is independent of scale and independent of statistics and, unlike the BMI, would therefore be meaningful if calculated for a single captured Martian.

According to Figure 12.3, there is a neat relationship between mean Rohrer index and age in children and some kind of equation could be fitted to the data for descriptive and predictive purposes. (Note, however, that Figure 12.1 gives no indication of scatter, so that we cannot gauge the reliability of predictions that might be made using it.) I point this out because the resulting equation would be inconsistent in terms of unit analysis: a unit-free index of body build would be related to age measured in units of time. This is one of countless situations where unit consistency cannot be expected to matter and where one should not be troubled by it. Whether the Rohrer index is better related to some other variable that correlates with age is another question.

12.3 The body mass index; estimating body fat from body mass and height

$$\text{Body mass index (BMI)} = \frac{\text{body mass, kg}}{(\text{height, m})^2}$$

The body mass index, or BMI, is well-known to the general public as an indicator of fatness. It is calculated as (body mass)/height^2. Its unit, not routinely mentioned, is usually now kg m^{-2}, although non-metric versions have been used.* This is an important topic here because unit analysis suggests that there is something amiss. Body 'fatness', expressed as percentage body fat, has no unit, so an index to be used as a measure of that might be expected to have no unit either. What would seem to make more sense would be 'mass per unit volume', with the unit being kg m^{-3}. This is because, divided by density (in kg m^{-3}), it would become 'volume per unit volume' and thus unit-free, just like percentage body fat. We have already seen that (body mass)/height^3, the Rohrer index, is a meaningful index of body build, so how do we make sense of the fact that the height exponent of the BMI is 2 rather than 3? We will come to the answer shortly, but it is useful to look first at other aspects of the BMI.

Because of what follows, it must be stressed at the outset that the BMI does work usefully as a measure of percentage body fat. And so it should. If someone lays down fat, then fat content and mass increase together, generally without change in height. It is also obvious that if two people of different height weigh the same, the shorter one is likely to have a greater relative fat content (though could instead be more muscular). An index of relative fat content should therefore have the properties of increasing with mass and, for a given mass, of decreasing with height, just as is true of the BMI. To that extent, at least, the BMI does make sense. Figure 12.4 shows that there is in fact a marked correlation between percentage body fat and BMI, the actual relationship differing between the sexes. Figure 12.4 also shows that there is considerable scatter in the relationships, so that the BMI is hardly reliable for judging the fatness of individuals. The following calculation strikingly reinforces this point.

Consider two individuals that differ in size, but share identical physical proportions and identical percentage fat contents, muscularity and density. For them the ratio of mass/height^3 has to be identical too. The BMI therefore cannot be identical in the two individuals, even though the percentage fat content is the same. This is because the heights differ. Now suppose that one of these individuals is 20 per cent taller than the other and that the smaller one has a BMI of 34 kg m^{-2}. As explained below, the BMI of the larger person must be 20 per cent higher, i.e. 40.8 kg m^{-2}. The World Health Organization guidelines categorize the smaller person as 'mildly obese' (class I) and the other as 'severely obese' (class III). (The intervening category corresponds to indices of 35–39.9 kg m^{-2}.)

So here we have two individuals of equal fat content being classified very differently. For better impact, some stages in the calculation were omitted, so let us spell out the reasoning in more detail. This time let us specify the smaller individual as

---

* Following my policy of deliberate inconsistency, I now write kg m^{-2} instead of kg/m^2.

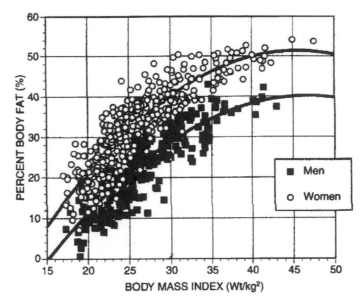

Figure 12.4   The relationship between percentage body fat and BMI in men and women. From Jackson *et al.* (2002), reprinted by permission from Macmillan Publishers Ltd (*International Journal of Obesity*, **26**, 789–796).

being of height 1.5 m and mass 76.5 kg. Those values are chosen to give the same BMI as above. Thus

$$\frac{76.5 \text{ kg}}{(1.5 \text{ m})^2} = 34 \text{ kg m}^{-2}.$$

The height of the larger individual is 20 per cent greater, and is therefore 1.2×1.5 m $= 1.8$ m. The volume of the larger individual is greater by the same factor cubed, namely by a factor of $(1.2)^3$, or 1.728. Since both individuals are postulated as having the same density, the larger mass is also bigger by this factor. It is therefore

$$1.728 \times 76.5 \text{ kg} = 132.2 \text{ kg}.$$

The BMI of the larger person is therefore

$$\frac{132.2 \text{ kg}}{(1.8 \text{ m})^2} = 40.8 \text{ kg m}^{-2} \text{ (as previously stated)}.$$

This example is a good illustration of the fact that the BMI is only a rough guide to fat content. Unfortunately the precise numerical values for individuals are sometimes taken more seriously than they should be. Thus large athletic and non-obese men have been rejected from military service on grounds of high BMI. Where the BMI is more valuable is in studies of groups, e.g. in epidemiological studies, where the general trends and correlations are more important than individual values.

Why is the BMI calculated as mass/height^2, rather than mass/height^3?

The basic property of the BMI is generally regarded as this, that it is a measure of mass that is uncorrelated, or at least minimally correlated, with height. In itself that has nothing to do with fat, but on the reasonable assumptions that fat contributes to total body mass and that percentage body fat is uncorrelated with height, at least in adults, then the BMI should correlate with percentage body fat. The following explanation for why the BMI is calculated as mass/height^2, concentrates on relationships between mass and height in adults, ignoring the issue of fat content.

Figure 12.5 shows data for 500 males. Note first that mass and height are very definitely correlated, but that there is considerable scatter. In other words one

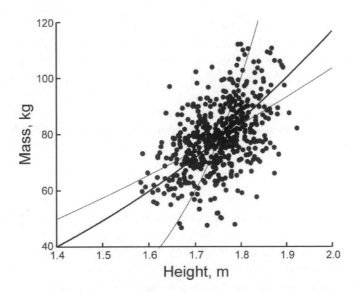

Figure 12.5 Simulated data on mass and height for a population of adult males. The thicker curve shows the true average relationship known to exist between mass and height. The shallower curve is for the regression of mass on height. The steeper curve is for the regression of height on mass.

might predict mass from height, or height from mass, but not reliably. Actually the data are not for real men, but have been generated artificially in a computer spreadsheet. The method was such that neither mass nor height is directly dependent on the other (see Notes). Some readers may have reservations about this use of simulated data, given the ready availability of real data. However, the point is that the true underlying relationship between the two variables is exactly known for the simulated data. This is never true of real data. More specifically, what is exactly known is the equation of the curve that may be drawn through the data and around which the individual data are randomly distributed. It is this:

$$\text{mass} = 14.6 \text{ height}^3. \tag{12.1}$$

The mean value of the Rohrer index, mass/height^3, is 14.6 kg m^{-3}. The means and standard deviations (scatter) for mass and height are typical of some populations of adult men (79 ± standard deviation 12 kg and 1.75 ± standard deviation 0.064 m respectively).

What about real data? A conventional belief is that mass varies more nearly with the square of height. This dates back to an assertion made by Quetelet in 1835, but it is backed up by much subsequent evidence. However, there is a fundamental difficulty with this evidence. If you look at the data in Figure 12.5 (ignoring the curves for the moment), you will appreciate that there is too much scatter for a curve to be fitted reliably by eye. The line of best fit must be calculated by some statistical technique and readers with any knowledge of statistics will probably think immediately of least-squares regression analysis (with the added complication that we do not have a straight-line relationship here). Figure 12.5 shows two regression lines – one for mass regressed on height and one for height regressed on mass. They cannot both represent the true underlying relationship, and in fact neither does. A similar point was made in Section 10.11 for a hypothetical straight-line relationship. There are other methods of curve fitting that may give better approximations to the true relationship, but the sad fact is that, without additional information, there is no certain way of establishing it (especially in this situation where neither variable determines the other). That is why Figure 12.5 shows simulated data; only that way can the true relationship be known.

It must not be thought that regression analysis is pointless here. The regression equations tell how best to estimate one of the variables from the other (a point discussed in Section 10.11). What is generally calculated is the nonlinear regression of mass on height, allowing mass to be estimated as reliably as possible from height. The equation here is:

$$\text{mass} = 25.0 \text{ height}^{2.05}. \tag{12.2}$$

This corresponds to the line with shallowest gradient in Figure 12.5. Note that the height exponent, 2.05, is close to 2. This is often true of real data, although values for different groups of adults have been found to range from about 1.1 to 2.5.

Therefore, by misunderstanding the nature of the statistics, one may conclude for the data of Figure 12.5 that mass is truly proportional, albeit approximately, to height squared, not height cubed. The data of Figure 12.5 were generated on the basis of a proportionality of mass and height^3. For reasons already explained, we do not know how true that is of real heterogeneous adult populations of the sort commonly studied. There does not seem to be much valid evidence on the matter. What is certain is that the usual supposition is wrong, namely that mass tends to vary with height squared.

In conclusion, there are two distinct indices here that are both valid in their way. As shown earlier, 'mass/height^3' is a valid measure of one particular aspect of body build. The BMI is a better index for predicting mass or fatness from height.

12.4 Upper arm muscle: how its cross-sectional area varies with body height

Much of the fat-free mass of the body is made up of muscle. Frisancho and Tracer (1987) studied the cross-sectional area of arm muscle in particular as a means of assessing nutritional status in children. Their data, including body height, are presented in tables in a way that is entirely appropriate to their practical aim. However, the data are worth re-examining with theory more in mind. Remember that, for constant shapes, areas vary with the square of linear dimensions. This applies to cross-sectional areas as well as to exposed surfaces.

The actual measurements in this study were of upper arm circumference, skinfold thickness over the triceps muscle, and body height. For purposes of calculation, the cross section of the arm was treated as circular and bordered by a uniform ring of adipose tissue with a thickness taken as half the skinfold thickness (because the 'pinch' of skin forming the skinfold consists of two layers). The area of the remaining central area could then be estimated by difference. The result was taken as the 'upper arm muscle area'. No correction was made for other tissues present with the skeletal muscle, but this should not matter in the comparison of children in different states of nutrition. Mean areas were tabulated in relation to body heights and those are the data we now look at.

What graph might one plot in order to interpret the relationship between upper arm muscle area and height? For consistency of units one should try first a plot of upper arm muscle area against height squared, rather than against either height or height cubed. Perhaps the data would then be suggestive of a straight line through the origin. Seeking a better rationale for doing the same thing, one might assume that the mass of upper arm muscle should be approximately proportional to the total mass of muscle in the whole body. Furthermore, if all the children of one sex had the same body build and relative muscle content, the total mass of muscle would be proportional to the fat-free mass and that, as we have seen, tends to vary with the cube of height, albeit inexactly. Finally in this chain of argument, since the data are for cross-sectional areas and not masses, it would be appropriate to plot them against height squared.

Figure 12.6 provides such a plot. The graphed points do lie roughly around straight lines through the origin, especially with the female data, but one would be right to be sceptical. However, perhaps one can see reasons for departures from linearity. Firstly, the heights of the youngest individuals are less than might be expected from their upper arm muscle areas. Does that not fit with the fact that these young children probably had relatively short legs? The upper arm muscle areas of the taller and older boys are bigger than corresponds to the postulated proportionality. Do not many boys develop more muscular arms at this stage?

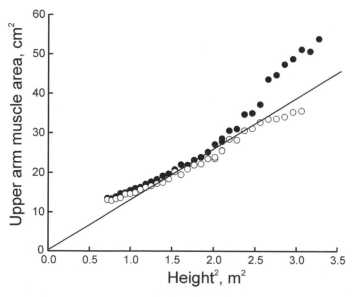

Figure 12.6 The relationship between the upper arm muscle area and the square of body height in boys (•) and girls (○). The data are those of Frisancho and Tracer (1987). The line through the origin is simply to guide the eye.

There are parallels in the interpretations of Figures 12.6 and 12.1. There is a general problem here that arises when one wishes to relate some variable to body size; the widest ranges are those associated with growth, but growth involves many changes other than size.

12.1.   Given data on both fat-free mass and upper arm muscle area for the same individuals, how might one plot them together in the hope that the data points would lie around a straight line through the origin?

12.5 Weightlifting – and the cross-sectional area of muscle

It is largely the cross-sectional area of a muscle that determines the maximum tension it can produce. This is understandable inasmuch as this area reflects the numbers of muscle fibres, and also, within these, the numbers of the myofilaments that run side by side in parallel. Dependence of force on cross-sectional area is best demonstrated with isolated skeletal muscle, but we can learn something simply by analysing weightlifting records. Table 12.1 shows records available to the author in 2007. The 'total lifted' refers to the sum of the weights lifted as 'snatch' and 'clean-and-jerk'. There is, for either sex, an obvious correlation between these totals and

Table 12.1   Weightlifting records for men and women, taken from *Guinness World Records 2007* (Glenday, 2007). What are shown are total weights lifted, arranged by weight class

Men		Women	
Class, kg	Total lifted, kg	Class, kg	Total lifted, kg
56	305	48	213
62	325	53	225
69	357	58	241
77	377	63	256
85	395	69	275
94	417	75	285
105	440	>75	305
>105	472		

the weight class* of the champion and it is this correlation that is analysed here. It is assumed that the cross-sectional areas of the relevant muscles are related to body mass.

Many muscles are involved in the coordinated act of weightlifting and their lengths alter during the lift. Therefore a full analysis of what is going on during a lift, including the lever actions of the bones, is complicated. However, one aspect of the matter can be treated very simply. In concentrating on the records for the various weight classes, we may make various simplifying assumptions. Firstly, it is a reasonable supposition that the champions of one sex all share something like the optimum body build, for otherwise they would not have become champions. Also we may suppose that they perform their lifts in similar fashion, and that, for a given sex, about the same proportion of the body is made up of skeletal muscle. This means that most of the potential sources of variability in the data are removed, other than body size, so that only a few data are needed for valid conclusions. If, by contrast, one were to investigate the weightlifting abilities of randomly selected individuals, one might require a very large sample size to show a significant relationship with body mass.

Body densities should be sufficiently similar in these champions that one may take the body volume as proportional to body mass (Chapter 11). As already discussed, areas for a given shape area proportional to volume$^{2/3}$. Thus, in terms of metres, $m^2 = (m^3)^{2/3}$. Then, given the assumptions made above, the cross-sectional area of all the relevant muscle can be taken as proportional to (body volume)$^{2/3}$ and so to (body mass)$^{2/3}$. If that is the case, then the maximum individual weight and total weight that each champion can lift should also be proportional to (body mass)$^{2/3}$.

The champions in Table 12.1 are classified by weight class. It would be useful to know their actual body masses, but we must be satisfied with the class specifications instead, bearing in mind that they are not ideal for our purpose. Unfortunately that will not do for the highest weight class because this has no upper limit, so the heaviest champions of each sex have to be disregarded.

Figure 12.7 shows total weight lifted plotted against weight class. The curves drawn through the points have the following formulae:

For men,   total weight lifted, $\text{kg} = 20.5 \times (\text{weight class, kg})^{2/3}$.   (12.3)

For women,   total weight lifted, $\text{kg} = 16.1 \times (\text{weight class, kg})^{2/3}$.   (12.4)

---

*A distinction should generally be made between weight and mass, with weight being equated to force. In this context of weightlifting that would seem to be unduly pernickety, for the things that weightlifters lift are called 'weights', and 'weight class' is a standard term. Besides, weightlifting is clearly about exerting force.

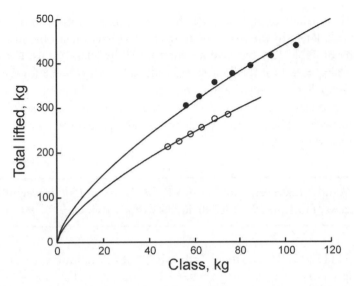

Figure 12.7 World records for total weight lifted plotted against weight class, both in kilograms. The equations of the curves, for men (•) and for women (○), are given in the text.

The data fit the curves remarkably well and this supports the hypothesis that, for a given sex, the total weight that each champion can lift is indeed proportional to (body mass)$^{2/3}$. However, the data do not quite fit the curves perfectly and other evidence suggests that the discrepancies are due mainly to a tendency for the heaviest lifters to have a slightly greater fat content (ca. 10 per cent). This implies that the relative muscle content is correspondingly less.

12.2 If Eq. (12.4) is valid for a 90 kg female champion, what total weight would she be expected to lift?

12.3 How might the data of Table 12.1 be plotted to obtain approximately straight-line relationships?

12.4 If the data for female weightlifters were shifted upwards in Figure 12.7, they could be made to lie close to the curve fitted to the males. It would be necessary to multiply all the female total weights by the same factor. What is that factor?

These relationships imply that the lighter weightlifters can lift more than the heavier ones relative to their own body masses. This is most obviously shown for the data of Table 12.1 by dividing 'total lifted' by 'class'. Thus, for women, (213 kg)/(48 kg) = 4.4 for the lightest and (285 kg)/(75 kg) = 3.8 for the nearly-heaviest; the smaller the body, the greater the cross-sectional area of muscle relative to body mass. According to Eq. (12.3), a tiny adult man of 1 kg, if such a person were possible, should be able to lift 20.5 kg. In that context the carrying abilities of ants seem less impressive.

> **12.5**   Would it make sense to abolish weight classes and judge weightlifters on the basis of total weight lifted divided by (body weight)$^{2/3}$?

Equations (12.3) and (12.4) may be satisfying, but what about further unit analysis? On the left of each equation we have kg; on the right we have $kg^{2/3}$. To ascribe the unit $kg^{1/3}$ to the numerical constants gets us nowhere, for what, measured in kilograms, is constant for each sex? In truth, those single numbers are empirical constants reflecting a complex situation. The lifted weights may be interpreted as representing forces rather than masses and therefore have force units (discussed in Chapter 14). At a given instant during a lift there are corresponding forces exerted by particular muscles that are proportional to their cross-sectional areas. For weightlifters of different sizes those areas may be taken as proportional to muscle mass raised to the two-thirds power and therefore to total body mass (or weight class) raised to the same power. For Eqs (12.3) and (12.4) to have unit consistency, units for the constants 20.5 and 16.1 must therefore be (unit of force)/$kg^{2/3}$.

12.6 Estimating body fat from skinfold thickness measurements

In this final example, we return to the estimation of body fat. A person's fatness is to some extent reflected in the thicknesses of skin that one can pinch between finger and thumb. Much more refined methods of measurement are used that involve the use of special measuring calipers at carefully selected sites. Various equations have been proposed for relating percentage body fat to skinfold thickness measurements. These are specific to particular populations or groups, defined by sex, age and race. The skinfold thickness measurements may be made at various sites and averaged to produce a single measure that I abbreviate here as STM without being more specific about it. Although the relationship between percentage body fat and STM may be expressed through a variety of equations, the basic idea is captured in the following simple version:

$$\text{Percentage body fat} = A \times \text{STM} + B, \qquad (12.5)$$

where A and B are group-specific constants. Other terms may be added that improve accuracy, but which are irrelevant to this account.

What do we learn by unit analysis? Percentage body fat has no unit; STM is measured in millimetres. This inconsistency suggests that the equation could be improved.

There is another reason for thinking that something important is missing from the equation. Is body size not relevant? A small person with the skinfold thicknesses of a larger individual would obviously carry proportionately more body fat. To consider an absurd extreme, a mouse with an ordinary human STM would be very obese indeed!

An equation that would solve the problem of unit inconsistency is this one, using fat mass instead of percentage fat:

$$\text{Fat mass} = C \times \text{STM} \times \text{height}^2 \times \text{fat density}, \qquad (12.6)$$

where C is a new unit-free constant. In terms of units, with STM and height expressed in centimetres, the two sides of the equation are consistent:

$$\text{kg} = \text{cm} \times \text{cm}^2 \times \text{kg}/\text{cm}^3 = \text{kg}.$$

If you are wondering why fat density is used, rather than body density, look at the result of dividing both sides of Eq. (12.6) by fat density:

$$\frac{\text{Fat mass}}{\text{fat density}} = \text{fat volume} = C \times \text{STM} \times \text{height}^2. \qquad (12.7)$$

Now both sides of the equation have units of volume.

Because the fat density may be taken as essentially constant, it may reasonably be combined with C to produce a new constant K (equal to $C \times$ fat density) as in the following simpler equation:

$$\text{Fat mass} = K \times \text{STM} \times \text{height}^2. \qquad (12.8)$$

Being equal to $C \times$ density, K has the same unit as density. There is some evidence that Eq. (12.8) does actually give better estimates of fat content than Eq. (12.5), but, at the time of writing, it does need to be tested on more data.

Although Eq. (12.8) is satisfactory in terms of unit analysis, one is left wondering why, in terms of causation or geometry, height should be squared. Perhaps it would be better to have height, not squared, but multiplied by some horizontal measurement instead. The latter might be expected to vary more or less in proportion to height, while being influenced also by some aspect of body build. As a simpler situation imagine an upright cylinder with a uniform covering of fat on the vertical sides. The volume of fat would be proportional to the product of its thickness, the height and some horizontal measure such as the circumference, diameter or radius. The body is more complicated, of course, for the subcutaneous adipose tissue is not even in thickness, and much of the body fat is away from the surface. Future study may reveal a suitable horizontal measurement.

12.6   Equation (12.5) gives estimates of fat content as percentages. Adapt Eq. (12.8) to do the same.

12.7 Postscript

A general principle underlies this chapter. For people or objects of the same shape and density masses are proportional to the cube of height (or to the product of three linear measurements) and areas are proportional to the $2/3$ power of mass. Random variations in shape make these relationships less exact, but the use of averages (as in Figures 12.1–12.3, 12.6 and 12.7) allows one to disregard some of this aspect of variation. Progressive variations of body shape with size do not invalidate the principle, but rather it is the very application of the principle that emphasizes any systematic variation in shape (with age, for example) and helps in its description.

This approach leads to an unconventional account of body form and composition, as well as to new ideas. It is less concerned than usual with the practicalities of prediction and estimation and is more about the understanding of relationships. It is also unusual in its general lack of statistics. Statistical analysis is generally a vital and pervasive element in this field of research, but this book is not about statistics and the question of statistical significance is not an issue here. Moreover, as I illustrate in regard to the BMI, regression analysis, so important in establishing prediction equations, can be unhelpful in elucidating the true underlying relationships that exist amongst variables.

Problems

12.1 See Section 12.4.

12.2 See Section 12.5.

12.3 See Section 12.5.

12.4 See Section 12.5.

12.5 See Section 12.5.

12.6 See Section 12.6.

Part IV

13

Introducing time

There has been little mention of time so far, or of related concepts such as frequency, speed and rate of flow. Their importance needs no emphasis, but the units even of force, pressure and energy can include time units despite their seemingly static nature. The latter are discussed in the next chapter. A few basic points need to be made about frequency, speed and acceleration, but this chapter is much more about the rates at which substances are carried in flowing fluids such as blood. Specific topics include the Fick Principle formula for estimating cardiac output and the concept of renal clearance. The Fick Principle is treated, not so much for its undoubted utility, as for the opportunity it offers for discussing ways of thinking about formulae in general.

Unlike the basic measures of Chapter 2 there are few alternative time units to trouble us. For social and astronomical reasons, time is resistant to decimalization (though France did try this briefly after the revolution). Indeed, scientists unconcerned with the highest levels of precision and accuracy are happy to use, not just seconds (s), hours (h) and minutes (min), but also days (d) and even years (y, yr or a) when the imprecision associated with leap years is unimportant. (The symbol 'a' is from Latin annus or French année.) Milliseconds (ms), microseconds (μs), nanoseconds (ns) etc. serve for shorter time intervals.

13.1 Frequency

Frequency is measured in reciprocals of time units. Cycles per second may be abbreviated as 'cps', but the name hertz (Hz) is preferred, with one cycle per second

Biomedical Calculations   Richard F. Burton
© 2008 John Wiley & Sons, Ltd

$(1\,\mathrm{s}^{-1})$ being 1 Hz. The hertz is named after Heinrich Hertz who discovered radio waves in 1888. Given that the resting heart beat of a resting athlete might be about $60\,\mathrm{min}^{-1}$, or $1\,\mathrm{s}^{-1}$, one could also say that its frequency is about 1 Hz. However, it is unusual to use hertz for such physiological phenomena. (It is probably unhelpful to remember, even for mnemonic purposes, that 'Herz' means 'heart' in German.)

The athlete's heart rate is written above as $60\,\mathrm{min}^{-1}$ or $1\,\mathrm{s}^{-1}$, but for unit analysis it can sometimes help to write 60 beats min^{-1} or 1 beat s^{-1}, specifying 'beats'. Thus, given that the amount of blood ejected from the left ventricle each beat is, say, 75 ml, one may choose to calculate the amount pumped per minute (the 'cardiac output') by writing explicitly:

$$60 \text{ beats/min} \times 75 \text{ mL/beat} = 4500 \text{ mL/min}.$$

'Beats' and 'beat' cancel out, but their presence makes the logic of the calculation clearer. In other contexts, words such as 'cycle' or 'event' may be appropriate.

In exactly the same way the ventilation rate, or expired minute volume, may be calculated as the tidal volume multiplied by the breathing frequency:

$$12 \text{ breaths/min} \times 500 \text{ mL/breath} = 6000 \text{ mL/min}.$$

13.1 In some of his recordings the saxophonist Charlie Parker improvised jazz at a rate of 660 notes per minute (and sometimes faster). How many milliseconds per note does that represent?

For comparison, the time elapsing between stimulation of the human motor cortex and consequent action potentials in the finger muscles is about 20 ms. A single beat of a hummingbird's wing may also take as little as 20 ms.

13.2 Speed and velocity

Speeds are measured in terms of distances per unit time. Thus a dragonfly or hummingbird might fly at 10 m s^{-1}, corresponding to 36 km h^{-1}. There is a distinction between speed and velocity. Thus, speed is the total path length moved

---

13.2 Tiny organisms seen under the microscope may seem to move about with great speed, but this is because distances are magnified while time is not. Depending on circumstances human spermatozoa can reach speeds approaching 60 μm s^{-1}, which is about one sperm length per second. What is that in cm h^{-1}?

per unit time regardless of direction, while velocity is a vector quantity, meaning that it depends on direction as well as magnitude.

13.3 Acceleration

Acceleration is a rate of change of speed or velocity and is calculated as the change divided by the time taken for the change. Its units therefore take the form of distance/time^2. Thus, although one may speak descriptively of metres per second per second, the units are given as m/s^2 or m s^{-2}.

Of special importance in the next chapter is the acceleration due to gravity, that is to say the acceleration of a falling body that is not slowed by air resistance. This is abbreviated as 'g'. On Earth, this may be taken generally as 9.8 m s^{-2}, but it varies a little from place to place. (Thus, when Jean Richer took a pendulum clock in 1671 from Paris to Cayenne in what was then French Guiana, he found that it lost two and a half minutes a day – until he returned it to Paris.) For many biological purposes, g may be approximated as 10 m s^{-2}, with only 2 per cent difference from 9.8 m s^{-2}. That makes arithmetic easier, but the commonplace number '10' does not proclaim its identity in the same way as '9.8'. Even if only for that reason it is therefore preferable to remember '9.8'. There is an international standard value for g of 9.80665 m s^{-2}, but this is too precise a value for biological calculations.

13.3   An object is falling against negligible air resistance. In 1 second its speed increases from zero to 9.8 m s^{-1}. What is its speed 1 second later?

13.4 Rates of flow of substances carried in fluids

Rates at which substances are carried along in a flow of fluid have been touched on in Chapter 1. We proceed now by means of a short set of problems for the reader to tackle. Deliberately, they are not all actually concerned with fluid flow, but there is a special point to the sequence. Even though these calculations are easily worked out in the head, the working should be recorded in detail. Units should be shown throughout, with cancellations where appropriate. Doing this throughout may help in answering questions 13.9 and 13.10.

13.4   A woman takes 10 ml of cod liver oil per day. It contains 0.12 mg of vitamin A per millilitre. What is her daily dose of the vitamin in mg day^{-1}?

13.5 A 2 mL sample of blood plasma is found to contain sodium at a concentration of 3.4 mg mL^{-1}. How much sodium is there in the sample (in mg)?

13.6 A pair of human kidneys typically filters about 180 litres of fluid per day. Its composition is like that of blood plasma less the plasma proteins, so that the concentration of sodium in the filtrate is about 3.5 g L^{-1} (150 mmol L^{-1}). What is the approximate rate at which sodium is filtered in grams per day?

13.7 In a particular individual, the blood in the right ventricle contains oxygen at a concentration of 150 mL O_2/L blood. It flows to the lungs at a rate of 6 L/min. At what rate is oxygen carried towards the lungs?

13.8 A man breathes out 5 litres of air per minute. It contains 4 per cent carbon dioxide (i.e. 4 L CO_2/100 L air). At what average rate does he expire carbon dioxide?

13.9 In terms of the mathematics, which one of the above calculations differs most in kind from the others?

13.10 For the other four calculations make up a single general formula that could be used to calculate any of their answers.

Ask yourself (a) whether making up the formula has helped you, (b) whether the formula is worth committing to memory and (c) if it is not worth memorizing, is that because the act of thinking it through has made a formula redundant. (There are no 'correct' answers to these three questions.)

13.5 Thinking about a formula

Suppose a lecturer presents a class with an unfamiliar and perhaps complicated-looking equation. Fighting off feelings of dread, a student may dutifully copy it down and hope to be able to learn it just before the exam. Others may look it in the face and try to befriend it. Let us consider one particular formula and how it may be demystified. The chosen formula is one given in many textbooks of physiology, but it may actually be advantageous now if you have never met it before. It has to do with the measurement of cardiac output, i.e. the rate of flow of blood in the aorta.

This 'Fick Principle' formula makes possible the calculation of cardiac output (CO) from whole-body oxygen consumption ($\dot{V}_{O2}$) (see below), and the concentrations of oxygen in arterial blood ($[O_2]_a$) and mixed venous blood ($[O_2]_v$). (The oxygen content of venous blood varies from vein to vein, so it is the 'mixed venous blood' entering the right atrium that is relevant here.) The formula is:

$$CO = \frac{\dot{V}_{O2}}{([O_2]_a - [O_2]_v)}. \tag{13.1}$$

Unit analysis of the Fick Principle formula

Let us see first how unit analysis works out, using units conventional in this context. Cardiac output is usually measured in L min^{-1} and oxygen consumption is commonly expressed as mL O_2 min^{-1}. Concentrations of oxygen in blood may be in mL O_2 (L blood)$^{-1}$. The units for the right-hand side of the equation are therefore:

$$\frac{mL\ O_2}{min} \div \frac{mL\ O_2}{L\ blood} = \frac{mL\ O_2}{min} \times \frac{L\ blood}{mL\ O_2} = \frac{L\ blood}{min} = L\ blood\ min^{-1}.$$

The units thus work out as is appropriate for blood flow. This can be seen as a test both of the formula and of the compatibility of the chosen units.

Applying common sense

Next we can consider whether the formula makes logical sense. For a given value of $([O_2]_a - [O_2]_v)$ would cardiac output (blood supply to the tissues) have to vary in proportion to oxygen consumption? For a given rate of oxygen consumption, would a greater blood flow be required if a given volume of blood gave up less oxygen in passing through the tissues? Are such considerations sufficient to determine the form of the equation?

Checking with realistic values

As a further test of the formula we may try it out on typical textbook values of the various quantities. Commonly quoted values for a resting person (the physiologist's standard 70 kg man – see Notes) are: 5 L blood min^{-1} for cardiac output, 250 ml O_2 min^{-1} for oxygen consumption, 200 ml O_2 (L blood)$^{-1}$ for arterial blood and 150 ml O_2 (L blood)$^{-1}$ for mixed venous blood. (The volumes of oxygen are for standard temperature and pressure – see Chapter 17.) These quantities should normally be inserted into the equation with the units. However, since unit analysis has already been performed, it suffices here just to use the numbers, thus emphasizing that a different test is being carried out. The right-hand side is calculated as:

$$\frac{250}{200 - 150} = 5.$$

The value 5 L blood min^{-1} is exactly the chosen cardiac output. However, one should not be surprised if a different set of textbook values failed to check out so exactly, for textbook values are typically just round numbers. Incidentally, if one knows only three of the four quantities, the equation allows the fourth to be obtained without reference to textbooks.

Deriving the Fick Principle formula

The next question is whether one can derive the equation for oneself. For this purpose, one may think of the body in rather simple terms. There is arterial blood carrying oxygen to the tissues in general (i.e. excluding the lungs). Some of this oxygen is removed from the blood by those tissues at rate $\dot{V}_{O2}$, i.e. the rate of oxygen consumption. Venous blood carries the unused oxygen away from the tissues. The rate of blood flow away from the tissues is the same as the rate of flow to them, i.e. the cardiac output, CO. In accordance with the generalization arrived at earlier that the rate of oxygen flow is the product of the rate of blood flow and oxygen concentration, oxygen is carried to and from the tissues at respective rates of CO $\times$ $[O_2]_a$ and CO $\times$ $[O_2]_v$. It remains to note that the three rates are necessarily related; the rate at which the tissues extract oxygen equals the rate at which it flows to the tissues less the rate at which it leaves them in venous blood. Thus,

$$\dot{V}_{O2} = CO \times [O_2]_a - CO \times [O_2]_v = CO([O_2]_a - [O_2]_v).$$

Exchanging the two sides of the equation and dividing both by $([O_2]_a - [O_2]_v)$, we get the original formula:

$$CO = \frac{\dot{V}_{O2}}{([O_2]_a - [O_2]_v)}. \tag{13.1}$$

(This derivation of the formula may also be worded in terms of the pulmonary circulation. In that case one refers, not to oxygen consumption by the tissues, but to the uptake of oxygen in the lungs. Mixed venous blood corresponds to pulmonary arterial blood and systemic arterial blood corresponds to pulmonary venous blood.)

Checking the algebra of the derivation

Having derived the equation, one might wish to check one's algebra using the method that has already been illustrated. To do this one inserts arbitrary numbers into Eq. (13.1), ignoring units. The numbers do not have to be at all realistic for this purpose, but can be chosen for ease of calculation. As noted earlier, the numbers 0, 1 and 2 are best avoided. Let us choose, for no particular reason, CO $= 3$, $[O_2]_a = 7$ and $[O_2]_v = 5$ and then calculate $\{CO \times [O_2]_a - CO \times [O_2]_v\}$ as $\{3 \times 7 - 3 \times 5\} = 6$. Thus $\dot{V}_{O2}$ works out at 6. All that remains is to see if these numbers are consistent when inserted into the final version of the formula. Thus

$$CO = \frac{\dot{V}_{O2}}{([O_2]_a - [O_2]_v)} = \frac{6}{(7-5)} = 3, \text{ the number originally chosen.}$$

It is unsurprising that the check works out satisfactorily in this case, but the trick can prove its worth more convincingly when applied to more complicated algebraic working.

Generalizing the formula to other situations

Equation (13.1), and the underlying principle, may be generalized to a variety of situations. What is required is that something ('X', usually a substance) moves at rate R_1 to a region (e.g. a tissue or organ), where it is removed or added to at rate R_2. Simultaneously X leaves the region by another route at rate R_3. Then

$$R_2 = R_1 - R_3, \tag{13.2}$$

where R_2 may be positive or negative. From this relationship any one rate is calculable from the other two. Both the measurements and the calculations are easier if either R_1 or R_3 can be taken as zero.

In which of these situations might the principle be applied?

(1) estimating the heat output from a domestic radiator from measurements of water flow and the temperatures of the water entering and leaving;

(2) estimating cardiac output from the average rate at which carbon dioxide is expired and the concentrations of carbon dioxide in arterial and mixed venous blood;

(3) estimating the oxygen consumption of an organ from the rate of blood flow through it and the concentrations of oxygen in the blood entering and leaving it;

(4) estimating glucose uptake by an organ from the rate of blood flow through it and the concentrations of glucose in the blood entering and leaving it;

(5) estimating heat loss from a water bird's feet from measurements of blood flow and the temperatures of the blood in the relevant arteries and veins;

(6) estimating a river's flow rate from the rate at which a pollutant is added to it by a factory and the concentration of that pollutant in well-mixed water downstream.

Estimating the rate of oxygen consumption

Equation (13.1) contains a term for the body's rate of oxygen consumption. This may be calculated from measurements made on expired air, this being collected, perhaps, in a Douglas bag or spirometer over a timed period of many breaths. The rate at which oxygen is expired is calculated from measurements of the volume, time and oxygen content. The rate at which oxygen is inspired is calculated from the same overall flow rate and the known oxygen content of air (21 per cent in dry air). The rate of oxygen consumption is then calculated as the difference between the two rates [cf. Eq. (13.2)].

So as not to obscure the point that the same principle is being applied as lies behind Eq. (13.1), the previous paragraph omits certain practical complications that need not hold us back here (see Chapter 17). Thus, air volumes vary with the temperature and pressure at which they are measured and also with their content of water vapour. Another complicating detail is that, even when air volumes are expressed in terms of dry air at standard temperature and pressure, there is usually a small differences between volumes expired and volumes inspired. This is because the oxygen removed in the lungs is replaced by a lesser volume of carbon dioxide. Typically the 'respiratory exchange ratio', the ratio of carbon dioxide produced to oxygen consumed, is about 0.8.

13.6 The concept of renal clearance

A well-known formula in renal physiology is that for 'renal clearance'. It relates particularly to the measurement of glomerular filtration rate (GFR) and renal plasma flow (RPF). Here it is:

$$\text{Clearance} = \frac{UV}{P},\qquad(13.3)$$

where V is the rate of urine flow, U is the concentration of a particular substance in the urine and P is the concentration of the same substance in the blood plasma (in the same units). UV is the rate at which the substance is excreted. (Note that the letter V that is conventionally used here, does not refer to a volume as it so often does, but to a volume per unit time. Some people therefore use $\dot{V}$ instead.)

Looking at the right side of the equation with the eyes of someone attuned to the idea of unit analysis, one sees that the units for clearance must be those of V, that is to say those appropriate to the rate of fluid flow. To spell this out, suppose that U and P each have the units mg/L and that V is expressed in mL/min. Then the right side is analysed as follows, to give the units of clearance:

$$\frac{\text{mg/L} \times \text{mL/min}}{\text{mg/L}} = \text{mL/min}.$$

Thus clearance is here in mL/min, so it is a rate of fluid flow just as GFR and RPF are. It is a striking fact that students study and revise this topic without registering this basic idea.

To measure GFR in terms of clearance one uses a substance that is readily filtered through the glomerular capillaries, but is neither reabsorbed nor secreted by the tubules, so that the concentration in the filtrate is the same as in the plasma (i.e. P). A suitable substance is inulin. Being a plant polysaccharide, it is not naturally present in the blood and has to be infused, but it has the essential qualities of being neither toxic nor bound to plasma proteins. Inulin has been used since 1935, but other substances are now in use, such iothalamate and Cr-EDTA. When labelled respectively with ^{125}I and ^{51}Cr, these may be measured in terms of their radioactivity.

13.11   Estimate the GFR from the following data.
Concentration of inulin in arterial plasma $= 122$ mg/100 ml
Volume of urine collected over 10 min $= 57$ ml
Concentration of inulin in urine sample $= 2235$ mg/100 ml

The GFR of a 'textbook' normal man is about 125 ml/min. Values in women tend to be about 10 per cent lower, even allowing for differences in size. For easy comparability, GFRs are sometimes expressed ('normalized') in terms of body surface area, calculated as in Chapter 11, the units being ml/min per 1.73 m^2.

To measure RPF, a substance with different properties is needed: ideally it should be completely cleared from the plasma as it flows through the kidneys. Suitable substances include *para*-aminohippuric acid (PAH), diodrast (iodopyracet) and hippuran because these are all both filtered and secreted into the urine by the tubules. (These can be measured in terms either of their iodine content, or of radioactivity if the iodine is labelled with ^{131}I.) Ideally the plasma should be completely cleared of the substance as it passes through the kidneys, but actually about 10 per cent escapes excretion, mainly because some of the blood passes through parts of the kidneys that are not active in excretion. What is measured is therefore described as the *effective* renal plasma flow (ERPF).

To reveal the logic of the clearance concept, Eq. (13.3) may be multiplied on both sides by P. Then

$$\text{Clearance} \times P = UV/P \times P = UV.$$

In the particular case of GFR, we have

$$\text{GFR} \times P = UV. \tag{13.4}$$

Here, yet again, we have rates of fluid flow multiplied by concentrations of substance in flowing fluids and that is true of both sides of the equation. Therefore Eq. (13.4) is telling us is that the rate at which inulin is filtered by the glomeruli equals the rate at which it is excreted. That we know to be the case, because inulin was chosen for that property. Perhaps the argument is best appreciated if it runs in the opposite direction, but to avoid repetition we use the reversed argument in looking now at renal blood flow.

The starting point is the fact that certain substances, such as PAH and diodrast, are almost entirely cleared from the blood as it flows through the kidneys. So the rate at which it reaches the kidneys is almost the same as the rate at which it is excreted. The rate at which it is delivered to the kidneys equals the rate of flow of blood plasma multiplied by the concentration in the plasma (once again a flow rate multiplied by a concentration). Thus,

$$\text{RPF} \times \text{concentration in plasma} = \text{rate of excretion.}$$

Therefore

$$\text{RPF} = \frac{\text{rate of excretion}}{\text{concentration in plasma } (P)}.$$

Similarly,

$$\text{rate of excretion} = \text{rate of urine flow} \times \text{concentration in urine}$$

$$= VU.$$

Therefore, putting the two equations together,

$$\text{RPF} = UV/P. \tag{13.5}$$

As already noted, this is really the 'effective renal plasma flow', because some of the PAH or diodrast reaching the kidneys escapes excretion.

13.12   Calculate ERPF from the following data:
Concentration of diodrast iodine in arterial plasma $= 0.64$ mg/100 ml
Volume of urine collected over 10 min $= 85$ ml
Concentration of diodrast iodine in urine sample $= 49.6$ mg/100 ml

---

Blood flow is calculated from plasma flow by taking into account the haematocrit.

13.13   A patient has an effective renal plasma flow of 500 ml/min and a haematocrit of 42 per cent. What is the effective renal blood flow?

A typical textbook value for the RPF for a resting adult is 700 ml/min, with the renal blood flow being 1.2–1.3 L/min. The 'filtration fraction' is the ratio of the GFR to the RPF. It is normally 0.16–0.20.

13.14   Because very little protein is filtered through the glomeruli, its concentration is higher in the efferent arterioles than in the afferent arterioles. If the filtration fraction is 0.20, what is the fractional increase in concentration?

13.7 Relating the clearance formula for renal plasma flow to the Fick Principle

It was pointed out that the Fick Principle can be generalized to a variety of contexts. Let us apply it to renal plasma flow measured using, say, diodrast. Renal plasma flow can be substituted for cardiac output, the flow to the whole body. For oxygen consumption we substitute 'diodrast consumption', meaning the rate of diodrast excretion, UV. For oxygen concentrations we substitute the concentrations of diodrast, calling the arterial concentration P. The concentration in the venous blood leaving the kidneys is again assumed to be almost zero. Equation (13.1) then matches Eq. (13.3):

$$\text{Clearance} = \frac{UV}{P-0} = \frac{UV}{P}.$$

13.8 Creatinine clearance as a measure of GFR, and a convenient formula for estimating it

The above method of measuring GFR is inconvenient for routine use. A marker substance must be infused into the blood, its concentration must be allowed to stabilize as it mixes within the whole extracellular fluid and its concentration (which falls continuously as a result of excretion) must be determined at the appropriate time. For estimating GFR in a clinical context it is more convenient to make use of a substance already present in the plasma, namely creatinine. This is steadily released from skeletal muscle. Unfortunately, a fraction of the filtered creatinine is secreted (typically about 10 per cent, but sometimes much more). Although the creatinine clearance therefore overestimates GFR, it has been a valuable measure of GFR for over 80 years. For the specific case of creatinine, Eq. (13.3) becomes:

$$\text{creatinine clearance} = \frac{\text{rate of creatinine excretion}}{P_{cr}}, \qquad (13.6)$$

where P_{cr} is the concentration measured in plasma or serum.

Sometimes it is inconvenient or difficult to obtain urine samples from a patient for the determination of creatinine clearance. In that case it can be estimated, though less reliably, just from the plasma or serum concentration. A well-known empirical formula for predicting creatinine clearance from P_{cr}, is that of Cockcroft and Gault (1976). With P_{cr} determined in mg/100 ml (mg/dL) the formula for adult males is this:

$$\text{creatinine clearance (mL/min)} = \frac{(140 - \text{age})(\text{body mass, kg})}{72 \times P_{cr}}. \qquad (13.7)$$

For females the estimate should be lowered by 15 per cent.

13.15 The concentration of creatinine in serum from a 24-year-old man weighing 75.1 kg is 1.1 mg/100 mL. Estimate his creatinine clearance.

To understand this formula we may start by combining Eqs (13.6) and (13.7) to obtain the following:

rate of creatinine excretion (mg/min) $= (140 - \text{age})(\text{body mass, kg})/72.$
$$(13.8)$$

A necessary condition for Eq. (13.7) to work well is that the plasma concentration, P_{cr}, should be steady and that implies that the rate of excretion equals the rate of production. Equation (13.8) can therefore be re-written as:

$$\text{rate of creatinine production (mg/min)} = \left(\frac{140 - \text{age}}{72}\right)(\text{body mass, kg}).$$
$$(13.9)$$

This shows the production rate as proportional to body mass and that may be seen as an easily measured, though inexact, substitute for the mass of skeletal muscle. Creatinine excretion rates, and therefore production rates, are known to correlate with muscle mass (and are sometimes used to estimate it). As for the term in brackets, '(140 − age)/72', this is an empirical expression that allows for a general decline in creatinine production with age. (It is also the case that creatinine clearance, i.e. GFR, shows a strong tendency to fall with age, but that is not immediately relevant here.)

Equation (13.6) is invaluable for bedside use, even though it is too unreliable for scientific research. One reason for its inaccuracy is that total body mass (or rather a fixed fraction of it) is standing in for muscle mass; allowances should therefore be made for discrepancies in obese patients. In addition it is assumed that P_{cr} is remaining steady during the period of measurement, meaning that the rates of creatinine excretion and production remain equal. This may not be the case. The effect of age must also vary with the individual.

Problems

13.1   See Section 13.1.

13.2   See Section 13.2.

13.3   See Section 13.3.

13.4   See Section 13.4.

13.5   See Section 13.4.

13.6   See Section 13.4.

13.7   See Section 13.4.

13.8   See Section 13.4.

13.9   See Section 13.4.

13.10   See Section 13.4.

13.11   See Section 13.6.

13.12   See Section 13.6.

13.13   See Section 13.6.

13.14   See Section 13.6.

13.15   See Section 13.8.

13.16   An individual on a protein-rich diet produced 1170 mL of urine in a day, containing 1200 mmol of nitrogen (mainly as urea) and 36.5 mmol of sulphate. What were the average concentrations of each in mmol/L?

13.17   An individual on low-protein diet produced 385 mL of urine in a day. The average concentrations of nitrogen and sulphate were 668 and 15.2 mmol/L, respectively. Calculate the rates of excretion in mmol/day.

13.18   Given the following measurements, calculate the cardiac output: oxygen consumption $= 211$ mL O_2 min^{-1}; oxygen content of arterial blood $=$ 193 mL O_2 (L blood)$^{-1}$; oxygen content of mixed venous blood $= 147$ mL O_2 (L blood)$^{-1}$.

13.19   Given the following measurements, calculate the cardiac output: carbon dioxide production $= 203$ mL CO_2 min^{-1}; total carbon dioxide content of arterial blood $= 485$ mL CO_2 (L blood)$^{-1}$; total carbon dioxide content of mixed venous blood $= 525$ mL CO_2 (L blood)$^{-1}$.

13.20   The concentration of creatinine in serum from a 75-year-old man 1.75 m tall and weighing 72 kg is 1.8 mg/100 mL. (a) Estimate his creatinine clearance. (b) Estimate his rate of creatinine excretion in mg/min.

Part V

14

Force, pressure, energy, work and power

These words have clear and restricted meanings in science that contrast with much popular usage – according to which people work under pressure, power may equate with strength and, as Shakespeare wrote, a 'sceptre shows the force of temporal power'. 'Energy' is used outside of science in such mysterious and mystical ways as almost to have become a marker for claptrap. As for 'weight', should scientists care that the word is improperly used to mean 'mass', or is that mere pedantry? In his short story *The Truth about Pyecraft*, H.G. Wells wrote about the fattest club man in London. Thanks to a secret Indian recipe for Loss of Weight, he ends up floating like a gas-filled bladder (or like an astronaut in space, we might say today) – weightless indeed, but still as fat.

The intention here is not to provide a substitute for physics textbooks. For good reason, the latter usually use a consistent set of units (i.e. SI units), but, as already stressed, it is important to be able to cope with other units when they arise.

14.1 Force and weight

The SI unit of force is the newton, with symbol N. As to what 1 N feels like, it may be thought of as the force exerted by an apple resting on the hand, specifically one with a mass of about 100 g. (Any connection with Isaac Newton's legendary apple is purely coincidental.)

Biomedical Calculations   Richard F. Burton
© 2008 John Wiley & Sons, Ltd

The force on an object due to gravity, and imparted to what it may rest on, is its weight and this is proportional both to its mass and to the local strength of gravity. The latter is expressed as an acceleration, namely that of an object in free fall. The SI unit for that is m s^{-2}. Force may therefore be expressed in terms of mass multiplied by acceleration or, in units, kg $\times$ m s^{-2} = kg m s^{-2}. For convenience, this combination of units has been given the name 'newton'. Thus

$$1 \text{ N} = 1 \text{ kg m s}^{-2}. \tag{14.1}$$

This equivalence of force and the product of mass and acceleration is more general and fundamental than has been so far implied, as it defines the relationship amongst the net force of any sort that acts on an object, the acceleration that the force produces if unopposed and the object's mass.

Consider a mass of 1 kg that is subject to gravitational acceleration (g) at a typical earthly value of 9.8 m s^{-2}. Its weight is 1 kg $\times$ 9.8 m s^{-2} = 9.8 kg m s^{-2} = 9.8 N.

14.1　To return to the apple of *about* 100 g, what would its mass more accurately need to be to exert a force of 1 N on the hand? Take g as 9.8 m s^{-2}.

Some readers may be puzzled by the distinction I am making between mass and weight, for the two are not only confused in common speech, but often also in scientific publications. At the same time, it is well known that astronauts on the moon have reduced weights – without loss of substance (i.e. of mass). Strictly, a 'weight' should be given in units of force. However, it is often unhelpful to use newtons for the latter, but one does have the option of another unit of force, the 'kg-force' or 'kg-weight'. Strictly, this corresponds to the weight of 1 kg when g has the standard value of 9.80665 m s^{-2} – in other words to 9.80665 N. The Swedish name for this, the 'kilopond', is sometimes used, but its use in English is not generally encouraged. It has to be said that the confusion of mass and weight causes few problems in the biomedical and life sciences, but it can be disastrous for unit analysis.

Equation (14.1) is worth remembering, by rote if necessary, although it can be easily and reliably obtained as the product of mass and acceleration units if one remembers why that is appropriate. The fact that 1 N corresponds to 1 kg m s^{-2} will be important when we consider the units of energy and power. If there should be a problem in recalling whether the name 'newton' was given to kg m s^{-2} or g m s^{-2}, remember that the kilogram is the fundamental SI unit, not the gram.

An older alternative to the newton, rarely used now, is the 'dyne'. This non-SI unit corresponds to g cm s^{-2} and equals 10^{-5} N. Note that 10^{-5} is neither of the two standard factors of 10^{-3} and 10^{-6} that are used with SI units

$$1 \text{ newton } (1 \text{ N})$$
$$= 1 \text{ kg m s}^{-2}$$
$$= 10^5 \text{dynes}$$
$$\approx 102 \text{ g force}$$

14.2 Pressure

Pressure is force per unit area, as in 'pounds per square inch'. It is especially relevant to the cardiovascular and respiratory systems (in which context Chapter 17 deals with gas pressures). Appropriate SI units are $N \text{ m}^{-2}$. Because $N = \text{kg m s}^{-2}$, $N \text{ m}^{-2}$ can also be expressed as $\text{kg m s}^{-2} \times \text{m}^{-2} = \text{kg m}^{-1} \text{ s}^{-2}$. However, for most people, there is no particular value in memorizing that. For convenience, 1 N m^{-2} may be written as 1 Pa, where Pa stands for pascal in honour of Blaise Pascal (1623–1662) who, amongst many other achievements, demonstrated the effects of altitude on atmospheric pressure. The kilopascal (kPa) is more used in physiology than the pascal.

Pressure may also be expressed in terms of bars, atmospheres and mmHg. Atmospheric pressure varies, but a standard 'atmosphere' (1 atm) has a fixed value equal to 101.3 kN m^{-2} (101.3 kPa). For inexact use, this happens to be conveniently (or perhaps confusingly) close to 10^5 N m^{-2} (or 100 kPa). The bar is defined as 10^5 N m^{-2}, so that 1 atm is very close to 1 bar or 1000 millibars (mb).

$$1 \text{ N m}^{-2}$$
$$= 1 \text{ kg m}^{-1} \text{ s}^{-2},$$
$$= 1 \text{ Pa}$$
$$1 \text{ mmHg}$$
$$= 1 \text{ torr}$$
$$= 1/760 \text{ atm}$$
$$= 100/750 \text{ kPa}$$
$$= 0.1333 \text{ kPa}$$

Because mercury barometers and manometers have long been common laboratory instruments, pressures may also be specified in terms of millimetres of mercury (mmHg), notably in relation to blood pressures and other aspects of physiology. Sometimes 1 mmHg is described as 1 torr, with 'dorr' honouring Galileo's pupil and secretary Evangelista Torricelli (1608–1647) for his pioneering work on the barometer. For the conversion of units note that

$$1 \text{ atm} = 760 \text{ mmHg} = 760 \text{ torr}$$

and

$$750 \text{ mmHg} = 750 \text{ torr} = 100 \text{ kN m}^{-2} = 100 \text{ kPa}$$

In the contexts of blood pressures and respiratory gases, conversions are most often needed between kPa and mmHg. Here are the conversion ratios:

$$1 = \frac{0.1333 \text{ kPa}}{1 \text{ mmHg}} = \frac{7.50 \text{ mmHg}}{1 \text{ kPa}}.$$

You might wish to decide which two of these various numerical values would be most usefully remembered by you personally. You could, for example, choose the standard atmosphere in the appropriate unit and '7.50' as the factor for conversion of kPa to mmHg.

14.3 Columns of water, columns of blood

Let us look now at the relationship between the height of a column of fluid, h, and the pressure (P) at the bottom of the column. More exactly, let us take this as the difference in pressure between the top and bottom of the column and excluding the atmospheric pressure. (One may then speak of the pressure at the bottom as 'the pressure above atmospheric'.) We could be considering, for example, the pressure at the foot of a column of mercury or water in a glass tube, or else the pressure on a fish at depth h below the surface of a lake. The formula relating P to h is

$$P = \rho g h, \tag{14.2}$$

where ρ is the density of the fluid and g is the gravitational constant.

Equation (14.2) can be checked for unit consistency, with individual terms expressed in kg, m and s as appropriate. The units for the right-hand side are:

$$\frac{\text{kg}}{\text{m}^3} \times \frac{\text{m}}{\text{s}^2} \times \text{m} = \frac{\text{kg}}{\text{m s}^2} = \text{kg m}^{-1}\text{s}^{-2}$$

and those have already been established as equivalent to N m^{-2}, as is appropriate for pressure.

Pressures are sometimes specified in 'cm H_2O'. This makes practical sense in relation to water-filled apparatus where differences in water levels are conveniently measured. One may apply Eq. (14.2) to convert cm H_2O to other units, but the simplest conversion is to mmHg, because one merely has to take account of the different densities of the two fluids. Water has a density close to 1 kg/L and mercury has a density of 13.6 kg/L. A column of mercury therefore has the same pressure at the bottom as a column of water that is 13.6 times as tall. Therefore

$$1 = \frac{1.36 \text{ cm } H_2O}{1 \text{ mm Hg}} = \frac{1 \text{ mm Hg}}{1.36 \text{ cm } H_2O} = \frac{0.735 \text{ mm Hg}}{\text{cm } H_2O}.$$

14.2   Convert 1 atm to cm H_2O.

14.3   Starting with the previous answer, calculate to the nearest metre the depth of lake water at which the hydrostatic pressure due to the water alone equals the atmospheric pressure at the surface.

Arterial blood pressures are normally measured at the level of the heart. In a person standing upright, the arterial blood pressure in the feet is higher and this difference, due to gravity, may be calculated from the difference in height and the density of the blood. Blood specific gravities (d_{20}^{20}) are typically about 1.050–1.056 in women and 1.055–1.064 in men, depending mainly on the haematocrit, so we may take a typical density as 1.06 kg/L. Variations either side of this are less than 0.5 per cent and therefore of little consequence. Indeed, in the context of varying and pulsatile blood pressure one does not go far wrong in treating the blood as if it were water.

14.4   How much is the arterial pressure increased by the gravitational effect 120 cm below the heart? Take the blood density as 1.06 kg/L and give the answer in mmHg.

14.4 Osmotic pressure and colloid osmotic pressure (oncotic pressure)

Osmotic pressure is quantified in Chapter 9 just in terms of concentrations (e.g. osmolality). That is appropriate because it is rarely useful to use actual pressure units, such as mmHg and kPa. At 37 °C the relationship between concentration and pressure is as follows:

$$1 \text{ mosmol/kg water} = 19.3 \text{ mmHg}$$
$$= 2.57 \text{ kPa.}$$

14.5   Convert 290 mosmol/kg water (a typical value for most human body fluids) to both mmHg and kPa.

The osmotic pressure is therefore much higher than any of the hydrostatic pressures (e.g. blood pressures) within the body.

The colloid osmotic, or oncotic, pressure of the plasma is the contribution to the total osmotic pressure made by plasma proteins. Its value typically varies between about 20 and 35 mmHg, with a common textbook figure being 25 mmHg. Since the total osmotic pressure, due largely to salts, is about 5600 mmHg or 745 kPa (see calculation above), the contribution of the proteins is relatively minute, i.e. about 0.3–0.6 per cent. This is an important point to note, but one that is commonly overlooked. This does not mean that the colloid osmotic pressure can be ignored, however. It is highly significant in the context of the movement of water across capillary walls, where water and small molecules pass readily through and proteins are too large to do so. In this situation the total osmotic pressure might be, for example, 5600 mmHg on one side of the endothelium and 5570 mmHg on the other, with the difference between these, namely 30 mmHg, being the colloid osmotic pressure.

Fluid movements across capillary walls generally, and also in the glomeruli where renal filtration occurs, depend on the balance both of hydrostatic pressures and colloid osmotic pressures across the walls (the 'Starling hypothesis'). (The four pressures, two hydrostatic and two osmotic, are often described incorrectly as Starling *forces*.) Representative calculations involving these pressures as presented in textbooks are too simple to detain us here. However, it is worth noting, for mnemonic purposes, that the balance of pressures in the capillaries may favour the movement of fluid either way between plasma and interstitial fluid. This means that the typical colloid osmotic pressure (20–35 mmHg) is somewhat similar to the typical capillary blood pressure near heart level.

Recall that 1 mmol of solute per kg water exerts an osmotic pressure of 19.3 mmHg at body temperature. Note now that 19.3 mmHg is not far from being a typical plasma colloid osmotic pressure (20–35 mmHg). Therefore, a typical plasma protein concentration should be just over 1 mosmol/kg water. Can we relate this to the molecular weights of the various plasma proteins and to their concentrations in g/kg water? Unfortunately there are complications to such calculations (see Notes), so the following calculation gives only a very approximate answer.

14.6   The molecular mass of plasma albumin is 68 000. If plasma contained this at a concentration of 1 mmol/kg water, what would be the concentration in g/kg water?

Typical concentrations of protein in the plasma are 65–80 g/kg water, of which about 60 per cent is albumin. The plasma globulins have much higher molecular masses and therefore contribute less to the colloid osmotic pressure.

14.5 Energy and work

Full appreciation of the concept of energy comes through considering it in its various forms, including heat, chemical energy and kinetic energy. Nevertheless it can be simply defined as the capacity to do work. When work is done on an object, it gains energy. Work thus involves the transfer of energy. Accordingly, work and energy are measured in the same units. A key to many energy calculations is the principle of the conservation of energy, meaning that the sum of all forms of energy in a closed system is constant.

When an object is moved against a constant force, the work done on it (and energy used) equals the product of force and distance. This applies when a weight is lifted through a vertical distance. To work out appropriate SI units for energy and work, we may take either

$$N \times m = Nm$$

or, with the definition of a newton in mind,

$$kg\ m\ s^{-2} \times m = kg\ m^2\ s^{-2}.$$

We can check this out with other familiar formulae. According to Einstein's well-known equation, E equals mc^2, where E stands for energy, m for mass and c for the speed of light. As noted in Chapter 1, this leads to the same conclusion regarding appropriate units for energy:

$$kg \times (m\ s^{-1})^2 = kg\ m^2\ s^{-2}.$$

A more mundane if less generally familiar formula leads to this conclusion in the same way, namely that for the kinetic energy of a moving object. In this situation the energy is calculated as $^1/_2 mv^2$, where v is the object's velocity.

Because energy is such an important concept, it has its special (derived) unit. In the SI system the basic unit is the joule (J), where

$$1\,J = 1\,N\,m.$$

A feeling for the joule may be obtained by thinking again of a 102 g apple and then of the energy required to lift it 1 metre.

That 1 J needed to lift the apple is the work done on the apple. It does not include the larger amount of energy used in moving one's arm to accomplish the action. It should also be noted in passing that, although one can calculate the work needed to raise an individual person a given vertical distance as the product of weight (i.e. N, not kg!) and distance, the total work done by that individual in raising himself or herself that distance is much greater. The calculated work as a percentage of the total work is the 'efficiency'. Efficiencies of 25 per cent are rarely achieved in human activities.

Better known to the general public than the joule is the 'calorie' (cal). This is the amount of energy needed to heat 1 g of water by 1°C (from 14.5 to 15.5°C); 1 cal equals 4.186 J. Be aware that 1000 cal, i.e. 1 kcal, is sometimes called a kilogram calorie, a large calorie, or a Calorie spelt with a capital C. Conversions between calories and joules are common in relation to nutrition and exercise physiology, so the conversion ratios are conspicuous in this box:

$$1 = \frac{4.186\,J}{1\,cal} = \frac{0.239\,cal}{1\,J}.$$

Many people equate calories with food. Table 14.1 gives the energy contents of carbohydrates, protein and fat. It also shows the amount of oxygen required to oxidize each of these to carbon dioxide and water (plus urea in the case of protein) and the associated amounts of energy released per litre of oxygen. The respiratory quotient is the ratio of carbon dioxide released to oxygen utilized in moles or litres.

Other units of energy exist, including the erg, the kilowatt hour, the litre atmosphere and the British thermal unit. One erg is 10^{-7} J, enough to lift the apple an imperceptible 0.1 µm. The British thermal unit (BTU), equal to 1.0558×10^3 J,

Table 14.1   Energy contents of carbohydrates and of typical protein and fat. Below these are the amounts of oxygen needed for complete metabolism or combustion, with the figures for protein allowing for the loss of energy in nitrogenous excretory products. Also shown are the amounts of energy released per litre of oxygen and the respiratory quotient (the number of carbon dioxide molecules released divided by the number of oxygen molecules consumed)

	Glucose	Glycogen, starch	Protein	Fat
kcal/g	3.7	4.2	4.4	9.3
kJ/g	15.6	17.5	18.5	39.1
L oxygen/g	0.75	0.83	0.96	1.98
kcal/L oxygen	5.0	5.0	4.6	4.7
kJ/L oxygen	20.9	21.1	19.2	19.8
Respiratory quotient	1.00	1.00	0.81	0.71

may be found on old domestic gas bills. The other two appear in Problems 14.16–14.18.

14.6 Power

Power is the rate of doing work. Appropriate SI units are therefore J/s, but 1 J/s is also called the watt(W).

$$
\begin{array}{|l|}
\hline
\text{1 watt (W)} \\
= 1\,\text{J/s.} \\
\hline
\end{array}
$$

Some feeling for the magnitude of a watt is to be obtained by consideration of the power ratings of domestic appliances. Take the case of a 100 watt light bulb. How does its rate of energy use compare with a typical daily metabolic rate? We can convert 100 W to kcal/day as follows:

$$
100\,\text{W} \times \frac{1\,\text{J/s}}{1\,\text{W}} \times \frac{0.239\,\text{cal}}{1\,\text{J}} \times \frac{1\,\text{kcal}}{1000\,\text{cal}} \times \frac{86\,400\,\text{s}}{1\,\text{d}} = 2065\,\text{kcal/day.}
$$

The answer is within the range for typical basal metabolic rates. Thirty people sitting quietly in a lecture theatre would thus be roughly equivalent to a 3 kilowatt electric fire. Of course, one has no feeling for one's basal metabolic rate, whereas one is aware of the mild exertion of walking. A person walking at 2.8 mph on tarmac might be working, say, at about 270 W. So the extra power, above that postulated 100 W basal rate, would then be 170 W.

14.7   What is that 2065 kcal/day (100 W) in terms of grams of glucose per day?

14.7 An overview of units – from mass to pressure and power

A sequence of concepts and units has been traced from mass to force and weight, to energy and work, and then to power – with a side branch to pressure. Figure 14.1 summarizes this. It also shows the involvement of distance (m) and time (s) even in static concepts that seem unrelated to time, namely force, energy and pressure.

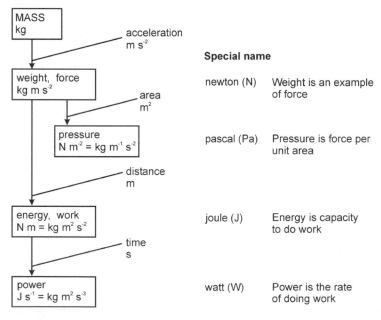

Figure 14.1   Summary of relationships amongst certain units, especially of force, energy and power.

Problems

14.1 See Section 14.1.

14.2 See Section 14.3.

14.3 See Section 14.3.

14.4 See Section 14.3.

14.5 See Section 14.4.

14.6 See Section 14.4.

14.7 See Section 14.6.

14.8 Convert 120 mmHg to kPa. (Why might that pressure have been chosen?)

14.9 Suppose that you need to estimate roughly the amount of heat needed to restore to normal the body temperature of an individual in hypothermia and that you know that person's mean body temperature and mass. Would you find it easier, or more appropriate, to work in terms of joules or calories?

14.10 One calorie = 4.186 J. In the context of physiological and nutritional measurements, do you think that a conversion factor of 4.2 would generally be accurate enough? Do you think that 4.186 or 4.19 might be easier for you to remember, as being more distinctive?

14.11 A student has a daily metabolic rate of 11 000 kJ/day. What is that in kcal/day?

14.12 According to the labelling on the packet, 100 g of cookies contain 43 g of sugar (3.7 kcal/g), 24.6 g of other carbohydrate (take as 4.2 kcal/g), 3.9 g of protein (4.4 kcal/g) and 15.9 g of fat (9.3 kcal/g). What is the energy content of 100 g of cookies in kcal?

14.13 If a man has a fairly typical resting metabolic rate of 1700 kcal/day, what is that in kJ/day?

14.14 A particular individual has a mean metabolic rate of 2500 kcal/day over a year and a net food intake (i.e. food minus faeces) of 2600 kcal/day (a 4 per cent difference that is equivalent to roughly 25 minutes walking per day). If all that excess energy were to go into fat, what mass of fat would be laid down over a year? The energy equivalent of fat is 9.3 kcal/g.

14.15 The typical oxygen consumption of a resting 70 kg man is often given as 250 mL O_2/min (as a round number). This corresponds to 360 L O_2/day. To check how well this squares with the 1700 kcal/day of question 14.13, calculate the ratio of kilocalories to litres of oxygen. Is it compatible with the data in Table 14.1?

14.16 The 'kilowatt hour' is a unit of energy. It represents an hour's worth of power at 1 kW. (Given the domestic context of the kilowatt hour, one may think of ten 100-watt light bulbs glowing for an hour.) Show its relationship to the joule, given that $1\,h = 3600\,s$.

14.17 The 'litre atmosphere' is another unit of energy. Imagine (or admit) that you have no idea of why this should be, or in what context it might be used. The name of the unit implies that energy or work can be calculated as a product of volume and pressure. Working in terms of metres and newtons, demonstrate that litre atmospheres can be re-expressed as newton metres. At this stage do not include numbers in your working.

14.18 Now work out a conversion factor for litre atmospheres and joules.

14.19 The evaporation of 1 g of water requires 2.43 kJ, the 'latent heat of evaporation' (see Notes). (a) How much water would need to evaporate to dissipate 100 J of heat from the body? (b) For an individual with a metabolic rate of 100 W, how much water would need to evaporate per hour (e.g. from sweat and the airways) if this were the only way of dissipating heat?

14.20 The free energy change (ΔG_{ATP}) for the hydrolysis of adenosine triphosphate (ATP) to adenosine diphosphate (ADP) is usually estimated for intracellular conditions as −10 to −13 kcal/mol of ATP (−42 to −54 kJ/mol). Much of our ATP is produced in the course of oxidative metabolism of glucose, with 1 mol (180 g) of glucose typically yielding 38 mol of ATP. (a) If the metabolism of 1 mol of glucose yields 38 mol of ATP and ΔG_{ATP} is 12 kcal/mol, how much energy is available from 1 mol of glucose by this route? (b) According to Table 14.1, 1 mol of glucose produces 666 kcal on complete combustion (i.e. 3.7 kcal/g × 180 g/mol). What percentage of this is available to the body via the hydrolysis of ATP? In other words, what is the efficiency of energy transfer from glucose to ATP?

15

Lessons from another formula

We have previously looked at various formulae, not necessarily for their usefulness, but rather as exercises in quantitative thinking and the application of unit analysis. For the same reason we look here at an equation which concerns the flow of fluids in cylindrical tubes. It is often quoted in relation to cardiovascular physiology, but more to summarize the various factors that immediately govern flow within a blood vessel than for use in real calculations. The equation was derived in 1838 by the physician and physiologist Jean Louis Poiseuille and independently by Gotthilf Hagen at almost the same time. Accordingly, the equation is called Poiseuille's equation or the Poiseuille–Hagen equation.

15.1 Poiseuille's equation and viscosity

The flow rate in a horizontal vessel or tube with a uniform radius is given by;

$$\text{flow rate} = \frac{\pi}{8} \times \frac{\text{radius}^4}{\text{length} \times \text{viscosity}} \times (P_1 - P_2), \tag{15.1}$$

where P_1 and P_2 are the hydrostatic pressures at the beginning and end of the tube respectively.* Particularly notable is the huge influence of radius, due to the

---

* If you are unsure how to make sense of such an equation, you might like to work through Problem 15.2 at this stage.

Biomedical Calculations   Richard F. Burton
© 2008 John Wiley & Sons, Ltd

fact that it is raised to the fourth power. Thus a halving of the radius reduces flow by a factor of $(1/2)^4$, i.e. 1/16. There are two important physiological implications to this. One is that it is the narrower blood vessels that contribute most to the peripheral resistance. The other is that arterioles, being both narrow and of adjustable diameter, are the most important vessels for controlling peripheral resistance and local blood flow.

Now we come to unit analysis. The term 'viscosity' in Eq. (15.1) may be unfamiliar, at least as regards the units in which it is measured. If you look this up elsewhere, you are likely to find that the unit given is the poise (obviously named after Poiseuille) or, more likely, the centipoise, because that has a magnitude more appropriate to many biological fluids. The viscosity of water at 20 °C is close to 1 centipoise. Motor oil has about 400 times the viscosity of water.

Equation (15.1) can only apply when all the units are appropriate. In using the equation quantitatively, one therefore needs to check the compatibility of the units. Presumably the 'mmHg' of physiologists will not do for unit analysis without the appropriate conversion factor – and, in this context, neither will the centipoise. However, the viscosity can also be expressed in SI units and our problem now is to work out what these SI units must be.

Rearrangement of Eq. (15.1) gives:

$$\text{viscosity} = \frac{\pi}{8} \times \text{radius}^4 \times \frac{1}{\text{length}} \times \text{pressure difference} \times \frac{1}{\text{flow rate}}.$$

From this we can work out the SI unit for viscosity:

$$\text{viscosity unit} = m^4 \times \frac{1}{m} \times \frac{N}{m^2} \times \frac{s}{m^3} = \frac{N\,s}{m^2} = N\,s\,m^{-2}.$$

These are therefore the appropriate viscosity units to use in Eq. (15.1). As for the poise, one of these equals 0.1 N s m^{-2}, so that a centipoise is 10^{-3} N s m^{-2}. Not many of us actually need to remember that, but the point of the exercise is to illustrate once again the use of unit analysis.

To make more intuitive sense of the viscosity unit N s m^{-2}, think of a viscous fluid being forced along a tube. The more viscous it is, the greater pressure difference is required to move it at a given rate and, for a given pressure difference, the longer time is needed for a given volume to flow. It should be no surprise that the viscosity unit is given by the product of pressure and time units (i.e. N m^{-2} × s = N s m^{-2}).

15.1   What is 'N s m^{-2}' re-expressed in terms of kilograms, metres and seconds?

It has to be said that there are difficulties in applying Poiseuille's equation to blood flow in an exact quantitative manner. Complicating factors include the pulsatile nature of the flow, turbulence and the fact that the viscosity of blood, which is a mixture of plasma and cells and not a uniform fluid, depends somewhat on the circumstances of measurement. In narrow vessels the blood cells tend to flow down the centre, away from the endothelial walls, except that in capillaries they may have to distort to squeeze through.

One may simplify Eq. (15.1) by writing:

$$\text{flow rate} = \frac{\text{pressured difference between ends}}{\text{resistance}} \qquad (15.2)$$

where the resistance is given by the reciprocal of the expression

$$\frac{\pi}{8} \times \frac{\text{radius}^4}{\text{length} \times \text{viscosity}}.$$

Then, if the analogy helps, you can relate Poiseuille's equation to Ohm's law:

$$\text{current flow} = \frac{\text{electrical potential difference (voltage)}}{\text{resistance}} \qquad (15.3)$$

Why is the vessel of Eq. (15.1) specified as horizontal? Consider a 100 mm column of mercury in a vertical test-tube. The pressure at the top (call it P_2) is that of the atmosphere. The pressure at the bottom (call it P_1) is that of the atmosphere plus 100 mmHg. Thus $(P_1 - P_2) = 100$ mmHg. Does the mercury flow vertically out of the tube away from where the pressure is highest? An extra (gravitational) term is needed in the equation if it is to accommodate flow (or in this case lack of it) between one level and another.

15.2 Peripheral resistance

So far this discussion has been about a simple tube, but the simplification of Eq. (15.2) may be adapted to the whole systemic circulation. Then the flow rate is represented by the cardiac output and the resistance is the 'total peripheral resistance'. The systemic blood vessels are not generally horizontal, of course, but this does not invalidate Eq. (15.2), provided that the pressure difference is taken as the difference in two pressures measured at the same level. The obvious level to choose is that of the heart.

The pressure difference, though varying during the cardiac cycle, may be taken just as the mean arterial pressure, because the venous pressure close to the heart

(P_2) is generally low enough to be taken as zero. We thus have:

$$\text{cardiac output} = \frac{\text{mean arterial pressure}}{\text{total peripheral resistance}}. \tag{15.4}$$

By rearrangement,

$$\text{total peripheral resistance} = \frac{\text{mean arterial pressure}}{\text{cardiac output}}. \tag{15.5}$$

The most convenient units for the resistance are those based directly on those in practical use for cardiac output and blood pressure, usually L min^{-1} and either mmHg or kPa, respectively. Then the resistance units are mmHg min L^{-1} or kPa min L^{-1}. Thus, for a mean arterial pressure of 100 mmHg and a cardiac output of 5 L min^{-1}, both being typical values, the total peripheral resistance is

$$\frac{100 \text{ mmHg}}{5 \text{ L min}^{-1}} = 20 \text{ mmHg min L}^{-1}.$$

Curiously and perhaps for some 'historical' reason or collective habit, one also still meets dyne s cm^{-5}, even when the values are calculated from mmHg and L min^{-1} (see 'Problems' for the conversion). In some papers the units are unspecified and unidentifiable.

Problems

15.1   See Section 15.1

15.2   Which of the following statements are true of Eq. (15.1)? (a) For a given flow rate, radius and pressure difference, the tube length and the viscosity would have to be inversely proportional to each other. (There is no implication of a causal connection between them!) b) For a positive flow rate, $P_1 > P_2$. (c) With all other terms held constant, the flow rate is inversely proportional to viscosity. (d) With all other terms held constant, the flow rate is inversely proportional to P_2.

15.3   Convert 1600 dyne s cm^{-5} to mmHg min L^{-1}, these being alternative units for peripheral resistance. An orderly approach is particularly appropriate here. (133.3 N/m^2 = 1 mmHg. 1 N = 10^5 dyne.)

16

Heat and temperature

Heat has been discussed as a form of energy and the calorie has been defined as the amount needed to raise the temperature of 1 g of water from 14.5 °C to 15.5 °C. The relationship between heat and temperature was clarified in about 1760 by Joseph Black, who showed, as a general law, that 'all bodies communicating freely with each other, and exposed to no inequality of external action, acquire the same temperature'. Black also clarified the concept of heat capacity, which we return to below.

16.1 Temperature scales

Figure 12.3 shows the relationship between the Celsius and Fahrenheit scales, treated in Problem 10.4 of Chapter 10 as an example of a straight-line relationship. The box below is a reminder of how to convert from one temperature scale to another. It may be best to memorize just one of these Celsius/Fahrenheit formulae, since it is too easy to mix them up.

Converting between Celsius (C) and Fahrenheit (F)

$$C = {}^5/_9(F - 32.0)$$
$$F = {}^9/_5 C + 32.0$$

Converting between Celsius (C) and Kelvin (K)

$$K = C + 273.15$$
$$C = K - 273.15$$

Biomedical Calculations   Richard F. Burton
© 2008 John Wiley & Sons, Ltd

Science uses the Celsius and Kelvin scales, which have the same size of unit (called a 'degree Celsius' or a 'kelvin'). The Fahrenheit scale has no particular virtues, except for familiarity in the USA and amongst older people elsewhere. The Celsius scale was formerly called 'centigrade', a name that was officially abolished in 1948 but which inevitably lingers on.

The temperature of a substance or system relates to the energy associated with the motions of its constituent particles, whether these be gas molecules or vibrating atoms in solids. Absolute zero is the temperature at which these motions cease.

On the *Celsius* scale water freezes, by definition, at 0 °C and boils, at sea level, at 100 °C.* The core temperature of the human body is about 37 °C.

On the *Fahrenheit* scale water freezes, by definition, at 32 °F and boils, at sea level, at 212 °F. The core temperature of the human body is about 98.6 °F.

The *kelvin* scale measures 'absolute temperatures' in that it starts at 'absolute zero' (0 K), which is −273.15 °C. It is defined in relation to the triple point of water (0.01 °C) at which temperature ice, water and water vapour are in equilibrium. For many biomedical purposes '273.15' is over-precise, and '273' will do. However, I use the more precise figure to keep the reader reminded of its value.

The Kelvin scale is especially appropriate in relation to processes and properties that vary proportionately with temperature. Absolute temperature, abbreviated T, features in various formulae later in the book – in relation to the thermal expansion of gases, for example. At about the time of writing a friend remarked that Tokyo, then at 36 °C, was twice as hot as Glasgow (the home of Joseph Black and Lord Kelvin) at 18 °C. Pressed on this point, he thought he would not have said this if the respective temperatures had been 2 °C and 1 °C (not that it would have been worth remarking on!). It is meaningful to say that the temperature is doubled if a gas, say, is heated from 200 to 400 K, as its heat content is likewise doubled and the motions of its molecules are twice as rapid. If, by contrast, the gas is heated from 18 to 36 °C, it is unhelpful to describe this as a doubling of temperature for, on the Kelvin scale, this is a change from 291.15 to 309.15 K, a rise of 6 per cent. The point is that expressing temperature differences as ratios only makes sense on a scale starting at absolute zero.

This is not, of course, to deny the scientific respectability of the Celsius scale in other contexts, such as in describing body and environmental temperatures. It is the difference between these two, not their ratio, that determines the rate at which heat is conducted between one and the other (a difference that is identical in kelvins and degrees Celsius). There are other important biological processes whose rates are much more dependent on temperature. Most notable of these are

---

* Celsius published his scale in 1742, but with the boiling point of water as zero and its freezing point as 100°. It was another Swede, the biologist Linnaeus, who inverted the scale to give our present version.

the responses of the physiological systems regulating body temperature, since a change in core temperature of a fraction of a degree may produce obvious visible effects. Here the effect is so steep as to suggest a phase change like that of a lipid or of ice to water.

16.2 The temperature coefficient, Q_{10}

A concept that is defined in terms of a temperature difference is the temperature coefficient or 'Q_{10}'. This quantifies the effects of temperature on the rates of physical processes, including chemical reactions and diffusion, and may also be applied to more complex biological processes, like metabolic rates. The Q_{10} may be calculated as the ratio of two rates measured at temperatures 10 °C apart. Its numerical value depends somewhat on the chosen temperature range (which is usually near room or body temperature) and the measure is used more because it is convenient and easily understood than for any theoretical basis. For most chemical reactions the Q_{10} is near 2–3, but it can be much higher.

If the rates of a particular process are not available for temperatures ten degrees apart, the following formula may be used. R_1 and R_2 are the rates at temperatures T_1 and T_2.

$$Q_{10} = \left(\frac{R_2}{R_1} \right)^{\left(\frac{10}{T_2 - T_1} \right)}.$$

16.1 For the situation in which T_1 and T_2 are $10°$ apart, show that this formula gives the same result as the simple method of calculation described previously.

16.3 Heat capacity and specific heat

In general and over limited ranges, the amount of heat added to a material and the associated rise in temperature are proportional to each other. The constant of proportionality is called the material's heat capacity. It follows that, for a given change in heat content, a large heat capacity is associated with a small change in temperature. Of course, the amount of material is relevant too, and it is usually more useful to think in terms of the heat capacity per unit mass, which is called the specific heat. The specific heat of a substance is the energy needed to raise the temperature of 1 g or 1 kg by 1 °C or 1 K. From the definition of the calorie given above, the specific heat of water is 1 cal/g °C (4.186 J g^{-1} K^{-1}). This is one of the highest values for common materials and is, for example, about 10 times as high as the specific heats of copper and glass. This is of great biological importance, as it stabilizes the temperatures of ponds and oceans – and also that of the human body with its high water content. The specific heat of the body is 0.8 cal/g °C.

Problems

16.1   See Section 16.2.

16.2   At a core body temperature of 37.0 °C, what is 0.1° as a percentage of the absolute temperature? (Think of the answer in the context of the exactness of temperature regulation.)

16.3   (a) How many kilocalories are needed to raise the temperature of an 80 kg person by 1 °C? Assume a specific heat of 0.8 kcal/kg °C. (b) Assuming that that person's metabolic rate happens to be 128 kcal/h and that all heat exchange with the surroundings is prevented, calculate the rate at which the body temperature rises.

16.4   A woman weighing 65 kg has a mean body temperature of 37.0 °C. She drinks 650 ml (650 g) of water at 20 °C. By how much would her temperature fall? Her specific heat is 0.8 cal/g °C. That of the water is 1.0 cal/g °C. Assume that there is no homeostatic response, so that her matching rates of heat production and loss remain unchanged.

16.5   Maximum shortening velocities of rat soleus muscle have been found to average 7.0, 3.6, 2.4 and 0.75 muscle fibre lengths per second at temperatures of 35, 25, 20 and 10 °C, respectively (Ranatunga, 1984). What is the Q_{10} for 25–35 °C, for 10–20 °C and for 20–25 °C?

16.6   What is the Q_{10} of a hypothetical process that proceeds at a rate proportional to absolute temperature? (Calculate it for a temperature range 5° either side of 25 °C.)

16.7   A hospital needs to purchase an electric shower unit. The power (in watts) must suffice to raise the water temperature by 25 °C when the flow rate is a modest 5 litres per minute. Since the calorie is defined in terms of the energy needed to heat water, it makes sense to start by thinking in terms of calories. (a) What is the necessary heat input in cal s^{-1}? This is to be equated with electrical power in watts. Remember that $1\,W = 1\,J\,s^{-1}$ and that $1\,cal = 4.186\,J$. (b) What is the previous answer in watts? The shower unit will need to have at least this power output.

17

Gases: dry and wet gas mixtures, partial pressures, gases in solution

Calculations to do with the composition of gas mixtures and pressures are particularly important for respiratory physiology – in relation to medicine, diving, aviation, mountaineering, etc. Special saline solutions for cells and tissues often need to be equilibrated with gases of specified composition. The calculations in this chapter are, as usual, conceptually and arithmetically easy, but one does sometimes have to think carefully when dealing with gas mixtures quantitatively. This is mainly because gas volumes decrease under compression and, when unrestrained, expand on warming, but the content of water vapour may also have to be taken into account. It is often appropriate, therefore, to express composition in terms of standard conditions of temperature, pressure and humidity. It is easiest to begin this topic by treating dry gases and gas mixtures, because the presence of water vapour can make for complications. Thus, if one collects moist air exhaled near body temperature, its composition will change as it cools to room temperature and water condenses out of it.

Here, after recalling the units of pressure (Section 14.2), we look first at the gas laws relating total volumes to pressure and temperature and at the concept of partial pressure as applied to gas mixtures. The latter concept is important later when we come to gases in solution.

Biomedical Calculations   Richard F. Burton
© 2008 John Wiley & Sons, Ltd

17.1 A reminder of units

The basic SI unit of pressure is $N\,m^{-2}$, but in clinical and physiological contexts it is more usual to use kilopascals, where $1\,kPa = 1000\,N\,m^{-2}$. Dominating the older literature especially is the 'millimetre of mercury' (mmHg), also called the torr. It is therefore particularly important to remember either that $1\,mmHg = 100/750\,kPa = 0.1333\,kPa$ or that $1\,kPa = 7.50\,mmHg$. One also meets the bar and the atmosphere. With1 bar equal to 100 kPa and 1 atm equal to 101.3 kPa, these two are near enough the same for many biological purposes.

Because of its familiarity, temperature is usually given on the Celsius scale, but the absolute temperature, in kelvins, is required in applying the gas laws; 37 °C is $(37 + 273)\,K = 310\,K$.

17.2 Natural variations in atmospheric pressure

The full range of atmospheric pressures recorded worldwide at sea level is about 653 mmHg (870 millibars, 87 kPa) in the eye of a typhoon to 813 mmHg (1084 millibars, 108 kPa) in Siberia in 1968. The overall ratio is thus 1.25.

Air pressure also decreases with increasing altitude. Up to about 1000 m from sea level the decrease is nearly linear, such that at a height, h, in metres, it is reduced by a factor of $(1 - 0.000113\,h)$. For altitudes up to 7000 m the pressure is better described as decreasing with height by a factor of approximately $(1 - 1.16 \times 10^{-4}h + 4.4 \times 10^{-9}h^2)$. (Do not concern yourself with unit analysis here: these are empirical expressions for which unit analysis does not work!)

17.3 The gas laws

Consider a constant amount of dry gas that is subject to changes in pressure, temperature and volume, its 'amount' being specified in moles. The relationships amongst the three variables are most easily thought about first by taking them in pairs. In the seventeenth century, Robert Boyle showed that the volume is inversely proportional to pressure when the temperature is constant (Boyle's law). This means, for example, that if a quantity of gas is compressed to half its volume, its pressure is doubled. It also means that the product (pressure × volume) is constant. Later that century Amontons showed that pressure increases in proportion to temperature when the volume is constant. Around 1787 J.A. Charles showed that volume increases in proportion to temperature when the pressure is constant (Charles's law). These three 'gas laws' are combined in the following equation, in which P, V and T stand respectively for pressure, volume and *absolute* temperature, and in which n is the number of moles of gas. R is a constant that depends on the units used.

$$\frac{PV}{T} = nR. \tag{17.1}$$

R is the same for all gases and is known as the universal gas constant. The equation implies that equal volumes of gases (at a given temperature and pressure) contain equal numbers of molecules, a fact finally established by Cannizzaro in 1858. Gases may depart slightly from the relationship of Eq. (17.1) – to a degree of little consequence for biologists – and it is usual, therefore, to relate the equation to so-called 'ideal' gases (i.e. those that do fit the law).

We look at the gas constant in more detail below, but for the moment let us keep things as simple as possible. Consider a quantity of dry gas of volume V_1 at a pressure P_1 and temperature T_1. Conditions then change, giving new values V_2, P_2 and T_2 with no change in the number of moles, n. Equation (17.1) applies to both conditions so that:

$$\frac{P_1 V_1}{T_1} = \frac{P_2 V_2}{T_2}. \tag{17.2}$$

This formulation is a convenient one for treating changes in all three variables.

Because gas volumes vary with both temperature and pressure, it is usual to specify volumes at a standard pressure of one atmosphere and a standard temperature of 0 °C (273.15 K). This condition is called 'standard temperature and pressure' (STP) or 'normal temperature and pressure' (NTP).

As to the relationship between volume and the number of moles of the gas, there is a simple relationship to remember:

> At standard temperature and pressure
>
> one mole of gas occupies 22.4 litres.

17.4 A closer look at Eq. (17.1) and the universal gas constant, with attention to units

Before proceeding to more practical aspects of Eq. (17.1), we may look more closely at it, partly as an exercise in unit analysis. This subsection is not essential to what follows, but the gas constant is relevant in other contexts that have nothing to do with gases (Chapter 20).

We have met a product of pressure and volume ('PV') before (Problem 14.17), noting that the units equate to those of energy. Equation (17.1) may be rearranged as:

$$PV/n = RT. \tag{17.3}$$

In terms of units, we now have energy (PV) per mole of gas (n) being shown as proportional to absolute temperature, with R being a constant. This is what one might expect, given that the heat content (thermal energy) of a collection of gas molecules increases with temperature.

To focus on the gas constant, R, we may rearrange Eq. (17.1) in another way:

$$R = PV/nT. \tag{17.4}$$

The numerical value of R must depend on the units used, but unit analysis allows us to work out for ourselves what are valid combinations of units. Let us spell out the expression PV/nT in SI units and then simplify the result:

$$N/m^2 \times m^3 \times 1/mol \times 1/K = N\ m/(mol\ K)\ or\ N\ m\ mol^{-1}\ K^{-1}.$$

Recall now that 'N m' may be replaced by 'J'. The appropriate units for R are then J/(mol K) or J mol^{-1} K^{-1}. This simplification of the units, emphasizing energy (J), avoids the suggestion that forces (N) and distances (m) might be independently relevant, a thought that is intuitively unhelpful.

As to the numerical part of the gas constant, for this combination of units it is 8.314. Other combinations of units require different numbers. We thus have:

$$R = 8.314\ J\ mol^{-1}\ K^{-1},$$
$$R = 1.99\ cal\ mol^{-1}\ K^{-1},$$
$$R = 0.0821\ L\ atm\ mol^{-1}\ K^{-1},$$
$$R = 62.4\ L\ mmHg\ mol^{-1}\ K^{-1}.$$

Of these four equivalent versions of R, only the units of the last two explicitly include pressure (atm, mmHg). As an exercise and a check on consistency, let us relate the third to a re-arranged form of Eq. (17.1). We then have:

$$V = nRT/P \qquad\qquad (17.5)$$

with

$$R = 0.0821 \text{ L atm mol}^{-1} \text{ K}^{-1}.$$

Under conditions of standard temperature and pressure, $T = 273.15$ K and $P = 1$ atm. Combining this information, we can calculate the volume of one mole of gas at STP. Thus,

$$V = (1 \text{ mol}) \times (0.0821 \text{ L atm mol}^{-1} \text{ K}^{-1}) \times (273.15 \text{ K}) \div (1 \text{ atm})$$

$$= 22.4 \text{ L}.$$

This brings us back to what we already know (see the box above).

17.5 Treatment of gas mixtures – percentages

The composition of a gas mixture is usually given in terms of the percentages of its constituents – i.e. percentages by volume. Provided the latter do not include water vapour, these percentages are unaffected by volume changes relating to temperature and pressure. The normal composition of atmospheric air is given in Table 17.1. If there is water vapour present, it behaves like the other gases, except that cooling may cause some of it to condense out.

Table 17.1 Percentage composition of normal dry atmospheric air by volume

Nitrogen	78.09
Oxygen	20.95
Carbon dioxide	0.04
Other gases	0.92

17.6 Treatment of gas mixtures – partial pressures, tensions

The various constituents of a gas mixture contribute to the total pressure in proportion to their percentages. For example, if oxygen makes up 21 per cent of dry air and the total pressure is 1 atm, oxygen contributes $21/100 \times 1$ atm $= 0.21$ atm to the total. This is called the 'partial pressure' or 'tension' of the oxygen, symbolized as P_{O2}, PO_2. or pO_2. In millimetres of mercury this is $21/100 \times 760$ mmHg $= 160$ mmHg. In kilopascals it is $21/100 \times 101.3$ kPa $= 21.3$ kPa. The total pressure equals the sum of all the constituent partial pressures.

If a gas makes up A per cent of a gas mixture, its partial pressure (tension) $=$

$$\frac{A}{100} \times \text{total pressure.}$$

17.7 Water vapour pressure

When gas mixtures are saturated with water, i.e. equilibrated with liquid water, the partial pressure of water depends on the temperature. Abbreviated, PH_2O, this is also called the water vapour pressure, WVP. Table 17.2 gives water vapour pressures at temperatures in the biologically relevant range. Physiologists may usefully remember the value at 37 °C, namely 47.0 mmHg or 6.3 kPa, since this

Table 17.2 Water vapour pressures at temperatures between 0 and 40 °C

°C	mmHg	kPa
0	4.6	0.6
10	9.2	1.2
15	12.8	1.7
20	17.5	2.3
25	23.7	3.2
30	31.8	4.2
35	42.1	5.6
37	**47.0**	**6.3**
40	55.3	7.4

applies to air in the lungs. If the dry air of Table 17.1 becomes moistened, the percentages of oxygen etc. are correspondingly reduced.

17.8 'Standard temperature and pressure, dry'

As noted above it is often appropriate to express gas volumes as the values at STP. Textbooks of chemistry and physics that explain this typically concern themselves just with dry gases, saying nothing about moist gas mixtures. In contrast, physiologists and other biologists often deal with moist gases such as expired air. It may then be appropriate to specify gas volumes as 'STPD', i.e. 'standard temperature and pressure, dry'. A practical approach is to start by physically drying the gas sample. Otherwise, if the gas can be assumed to be saturated with water vapour, one simply deducts the water vapour pressure from the total pressure. Let us consider an example, working in kilopascals.

Suppose that one has collected 15.0 L of air at an ambient temperature and pressure of 15 °C and 98.0 kPa, respectively, and that it is saturated with water vapour. (It could be in a spirometer or a flaccid and impervious plastic bag like a Douglas bag.) The 15.0 L is to be converted to the value at STPD. From Table 17.2 the WVP corresponding to saturation at 15 °C is 1.7 kPa. Therefore the contribution of gases other than water vapour to the total pressure is $(98.0 - 1.7)$ kPa $= 96.3$ kPa. Suppose that the water vapour is removed with a drying agent, with consequent reduction in volume. The volume would be reduced by a factor of $96.3/98.0 = 0.983$. Thus the 15.0 L would become 15.0 L $\times$ 0.983 = 14.7 L. The dry gas is actually still at the ambient pressure of 98.0 kPa – instead of at the standard value of 101.3 kPa. At the higher pressure the volume would be less. A correction factor of 98.0 kPa/101.3 kPa = 0.967 therefore needs to be applied, reducing the volume to 14.7 L $\times$ 0.967 = 14.2 L. Next the temperature is lowered from 15 °C (288 K) to 0 °C (273 K) – with further shrinkage of the volume. The 14.2 L is therefore to be multiplied by (273 K)/(288 K), namely by 0.948. The volume at STPD would therefore be 0.948×14.2 L $= 13.5$ L. Reasoning that way it is easy to get the calculation wrong, and unit analysis cannot substitute for other thinking here. For repeated calculations it therefore helps to have a formula.

Equation (2) applies to dry gases. To make it apply to moist gases the water vapour pressure must be deducted from the total pressures:

$$\frac{(P - P H_2 O)_1 V_1}{T_1} = \frac{(P - P H_2 O)_2 V_2}{T_2}. \tag{17.6}$$

By rearrangement this becomes:

$$V_2 = V_1 \times \frac{(P - PH_2O)_1}{(P - PH_2O)_2} \times \frac{T_2}{T_1}. \tag{17.7}$$

This can be applied in a variety of circumstances, but we are looking specifically for a formula for converting gas volumes to STPD. Accordingly, let us make STPD correspond to condition 2, renaming P_2 as $P_{standard}$ and V_2 as V_{STPD}. Then, with $PH_2O = 0$:

$$V_{STPD} = V_1 \times \frac{(P - PH_2O)_1}{P_{standard}} \times \frac{273.15 \text{ K}}{T_1}. \tag{17.8}$$

The subscript 1 now refers specifically to the state of the gas sample before conversion to STPD. According to the pressure units used, $P_{standard}$ could have values of 1 atm, 760 mmHg, 101.3 kPa, etc.

17.1 Check the calculations of the previous paragraph using Eq. (17.8).

As well as 'STPD' (standard temperature and pressure, dry), there are other conditions distinguished by acronyms. One applies within the lungs, where the air is saturated with water vapour at body temperature (usually taken as 37 °C) and at a pressure virtually that of the ambient air. This condition is 'BTPS' – for 'body temperature and pressure, saturated'. Another condition, already met, is 'ATPS', in which the gas is saturated with water vapour at ambient temperature and pressure, as in a spirometer. A fourth is also for gas at ambient temperature and pressure, but dry (ATPD). Volumes under these various conditions may be specified as V_{ATPS}, V_{BTPS}, etc.

17.2 Adapt equation 17.7 specifically to calculate V_{BTPS} from V_{ATPS} when the atmospheric pressure is P_{amb} in mmHg, the ambient temperature is T_{amb} in degrees Celsius and the water vapour pressure at that temperature is w.

$$\frac{(P - PH_2O)_1 V_1}{T_1} = \frac{(P - PH_2O)_2 V_2}{T_2}.$$

17.9 Dissolved O_2 and CO_2 in blood plasma and other fluids

The concentration of a gas in simple solution at equilibrium with a gas mixture increases with the partial pressure (tension) of the gas in question (e.g. PCO_2, PO_2) and with the solubility coefficient (e.g. SCO_2, SO_2).* Thus, with square brackets denoting concentrations,

$$[\text{dissolved } CO_2] = SCO_2 \, PCO_2 \qquad\qquad (17.9)$$

and

$$[\text{dissolved } O_2] = SO_2 \, PO_2. \qquad\qquad (17.10)$$

Appropriate units for the solubility coefficients depend on those used for concentration and partial pressure. To take the case of oxygen, Eq. (17.10) may be rearranged as follows:

$$SO_2 = [\text{dissolved } O_2]/PO_2.$$

Units therefore include mmol O_2 L^{-1} $mmHg^{-1}$, mL O_2 L^{-1} kPa^{-1} and so on. Numerical values for the solubility coefficient depend on the gas in question and on the other components of the solution, including notably the salt concentration. In addition gas solubilities increase markedly with decreasing temperature. Thus, in plasma, SCO_2 is close to 0.03 mmol CO_2 L^{-1} $mmHg^{-1}$ at 37 °C, but 0.05 mmol CO_2 L^{-1} $mmHg^{-1}$ at 20 °C. For oxygen, more usually treated in terms of volume (at STP), SO_2 is 0.03 mL O_2 L^{-1} $mmHg^{-1}$ at 37 °C and 0.04 mL O_2 L^{-1} $mmHg^{-1}$ at 20 °C. When gas concentrations are expressed per litre of solution, proteins lower solubility coefficients just by taking up space.

Problems

17.1   See Section 17.8.

17.2   See Section 17.8.

---

* The symbol SO_2 is also used to denote the extent of oxygen saturation.

17.3 Some South Americans live at an altitude of about 5500 m (18 000 feet, 3.4 miles). What is the air pressure as a fraction of that at sea level? (see section 17.2)

17.4 Without knowing the question, calculate the answer using all of the following quantitative information. Then state the question. It concerns a gas-filled balloon. No other data are needed. $R = 0.0821$ L atm mol^{-1} K^{-1}; quantity of helium $= 0.2$ mol; pressure $= 1.02$ atm; temperature $= 300$ K.

17.5 Given that the molecular mass of oxygen is 32 (i.e. 1 mol O$_2$ = 32 g), what is the mass of 22.4 L of dry oxygen at STP? What is its density in g/L and kg/m^3?

17.6 If air at 760 mmHg contains 78.1 per cent nitrogen, what is the latter's partial pressure?

17.7 If the atmospheric pressure is 100 kPa and the air in the lungs is saturated at 37 °C, so that PH_2O is 6.3 kPa, what percentage of the alveolar air is water vapour?

17.8 In each of the examples in Table 17.3 convert the initial volume of gas, condition 1, to the volume corresponding to condition 2. Where applicable, classify each condition as STPD, BTPS or ATPS.

Table 17.3

		Condition 1			Condition 2		
	Volume	Pressure	°C	PH$_2$O	Pressure	°C	PH$_2$O
a.	4.00 L	96 kPa	20	2.3 kPa	101.3 kPa	0	0
b.	2.00 L	760 mmHg	0	0	760 mmHg	37	47 mmHg
c.	20.0 L	740 mmHg	15	12.8 mmHg	740 mmHg	37	47 mmHg
d.	20.0 L	752 mmHg	37	47 mmHg	760 mmHg	0	0
e.	3.00 L	95 kPa	37	6.3 kPa	95 kPa	15	1.7 kPa

17.9 A man breathes in dry air and breathes out moist air at a rate of 14 000 L/day expressed at STP (9.7 L/min). (a) What is that as moles of gas per day? (b) His expired air is saturated with water vapour at 33 °C, at which temperature the water vapour pressure is 37.7 mmHg. Atmospheric pressure is 759 mmHg. What is the percentage of water vapour in his expired air? (c) In grams, how much water does he lose daily in this way?

17.10 Given a normal human arterial PCO_2 of 40 mmHg and SCO_2 of 0.03 mmol/(L mmHg), what is the concentration of dissolved CO$_2$ in plasma in mmol/L? (Compare this with the concentration of bicarbonate, which is usually about 25 mmol/L and part of the 'total carbon dioxide'.)

17.11 Given an arterial PO_2 of 100 mmHg (approximating to normal) and SO_2 of 0.0013 mmol/(L mmHg), what is the concentration of dissolved O$_2$ in the plasma in mmol/L?

   The answer, about the same for whole blood as for plasma, may be compared with the total concentration of O$_2$ in fully oxygenated whole blood of about 9 mmol/L (equivalent to 200 mL/L).

17.12 If SCO_2 is 0.030 mmol CO$_2$ L^{-1} mmHg^{-1}, what is that in mL CO$_2$ L^{-1} kPa^{-1}? (Treat volumes as being at STP.)

Part VI

This final part contains all the topics, concepts, graphs and calculations that involve logarithms. These include Nernst equations, pH, the Henderson–Hasselbalch equation, dose–response curves and acid–base balance.

18

Introduction to logarithms

Before the invention of the pocket calculator, tables of logarithms were familiar to all students as an invaluable aid to multiplication, division and the calculation of roots. I learned to use them in this way long before I understood their meaning, but their nature is hardly mysterious. The definition may be summed up in a formula, but let us first look at some examples. The numbers 10, 1000 and 0.01 may be written respectively as 10^1, 10^3 and 10^{-2}. The logarithms of 10, 1000 and 0.01 are respectively 1, 3 and –2, these being the exponents of 10^1, 10^3 and 10^{-2}. More exactly, these should be referred to as 'common logarithms' or 'logarithms to the base 10', with 10 being the number that is raised to the indicated power. (We come later to 'natural logarithms', i.e. logarithms to the base e.) One should make a point of remembering that the logarithm of 1 is 0, which may be less obvious, but is true because $10^0 = 1$.

Neither these indices nor the logarithms are limited to whole numbers. Thus $10^{0.5}$, or $10^{1/2}$, is the square root of 10, namely 3.162, and 0.5 is the logarithm of 3.162. This statement can be written as '$\log_{10} 3.162 = 0.5$', or more simply as '$\log 3.162 = 0.5$'. The subscript 10 in '$\log_{10} 3.162$' indicates the base of the logarithm. Zero has no logarithm.

A calculator is invaluable in testing out the ideas presented here, so ensure that you know how to obtain logarithms with your model and can then convert the results back to the original numbers. For the latter, there may be a button marked '10^x'. Try obtaining log 0 on your calculator!

Biomedical Calculations   Richard F. Burton
© 2008 John Wiley & Sons, Ltd

18.1 Definitions

Generalizing from the above examples, we see that, if $A = 10^a$, $\log_{10} A = a$. One can also combine these and write:

$$\log_{10} (10^a) = a. \tag{18.1}$$

It is also true that

$$10^{\log a} = a. \tag{18.2}$$

The number to which a logarithm refers is called its *antilogarithm*, so that 10^a is the antilogarithm of a. The word was used more often when the values routinely had to be obtained from tables.

> If $A = 10^a$, $\log_{10} A = a$.

18.2 Rules for working with logarithms

Three rules need to be remembered:

$$\log (xy) = \log x + \log y \tag{18.3}$$
$$\log (x/y) = \log x - \log y \tag{18.4}$$
$$\log (x^y) = y \log x. \tag{18.5}$$

Taking x as equal to 1, we see that the following is a special case of Eq. (18.4):

$$\log (1/y) = - \log y. \tag{18.6}$$

18.3 The usefulness of remembering $\log_{10} 2$

To six places of decimal, the logarithm of 2 is memorably symmetrical, namely:

$$0.301030$$

(With 'punctuation' ignored, this is a palindrome, just like the bad waiters' slogan 'Stressed, no tips? Spit on desserts!') Because of its internal zeros, log 2 is conveniently approximated as 0.301 – or just 0.3 when less accuracy is required. Log_{10} 2 is worth remembering, along with log_{10} 1 (= 0) and log_{10} 10 (= 1), because several other logarithms are easily obtained from them without recourse to a calculator. This can be both useful and impressive to the uninitiated.

Moreover, it is useful practice to work through the derivations of other logarithms from log 2 and log 10, applying Eqs (18.3) and (18.4):

$$\log 4 = \log (2 \times 2) = 2 \log 2 = 2 \times 0.301 = 0.602.$$

$$\log 8 = \log (4 \times 2) = \log 4 + \log 2 = 0.903.$$

$$\log 5 = \log (10/2) = \log 10 - \log 2 = 1 - 0.301 = 0.699.$$

$$\log 2.5 = \log (5/2) = \log 5 - \log 2 = 0.699 - 0.301 = 0.398.$$

Note that these are very close to 0.600, 0.900, 0.700 and 0.400, respectively. One can derive yet more logarithms if one wishes:

$$\log 1.6 = \log (8 \times 2/10) = \log 8 + \log 2 - \log 10 = 1.204 - 1.000 = 0.204.$$

$$\log 3.2 = \log (1.6 \times 2) = \log 1.6 + \log 2 = 0.505.$$

Equation (18.5) provides an alternative derivation of log 8:

$$\log 8 = \log 2^3 = 3 \times \log 2 = 0.903.$$

18.4 Logarithmic scales on graphs

It is often useful to use logarithmic scales for one or both axes of a graph. This may be appropriate (1) when it matches the mathematical nature of the relationship being plotted (as illustrated in Chapter 19), (2) when the variables in question span too many orders of magnitude to be conveniently shown on ordinary linear scales or (3) again as a matter of convenience, when the plotted points happen to lie in a straight line. Figure 18.1 illustrates three ways of presenting logarithmic scales, i.e. of labelling the axes. The top version shows numbers increasing by factors of 10 from 1 to 10 000, but equally spaced. It is important to note that zero can never be shown on such a scale. Extension of this scale to the right soon leads to numbers that take up too much space (e.g. 1 000 000 000), so it is common to write them as 10^0, 10^1, etc., as in the second version. This has the added advantage just now of emphasizing the relationship with the bottom version, in which the numbers are replaced with their logarithms.

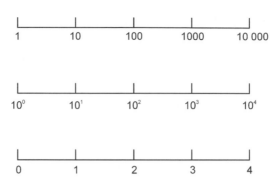

Figure 18.1 Three versions of a logarithmic x-axis scale. The upper two show actual numbers and the lower one shows them as their logarithms

Figure 18.2 shows a more limited scale, from 1 to 10, with ticks marking the positions of the integers from 1 to 10. Note their unequal spacing.

Graph paper is available that has logarithmic scales for one or both axes and the grid lines ruled accordingly. However, in the initial analysis of data it is generally best to calculate logarithms and use ordinary graph paper.

Figure 18.3 shows some data that have been chosen to span several orders of magnitude. The scales are not logarithmic and that may seem satisfactory – except that it is impossible to see that there are 20 data points for species with body masses less than 1 kg. These are evident in Figure 18.4, which shows the same data plotted using logarithmic scales. The line that passes through the origin in Figure 18.3 now has a gradient of 1.0, as is appropriate to simple proportionality. The data could also have been plotted as their logarithms on linear scales, with exactly the same effect. This method of plotting is illustrated in the next example.

18.1 Why does a gradient of 1.0 imply simple proportionality?

Figure 18.5 is a new version of Figure 12.7, showing weightlifting data from Section 12.5. However, it is now the logarithm of the weightlifting class that is

Figure 18.2 Logarithmic x-axis scale covering the range 1–10, with intermediate integers just shown as ticks

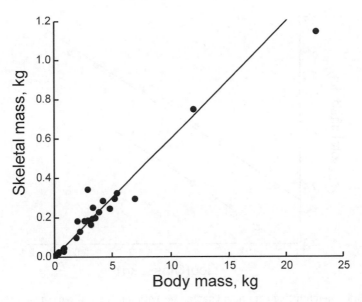

Figure 18.3 Body masses and dry skeletal masses (kg) of mammals ranging in size from shrew to dog. The gradient of the line is 0.06. The data are from Prange *et al.* (1979)

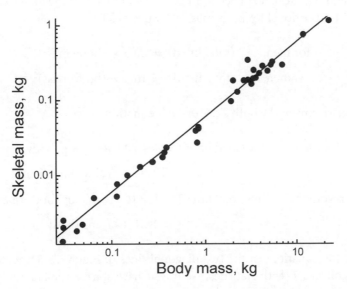

Figure 18.4 Skeletal masses plotted against body masses on logarithmic scales (data of Figure 18.3). The line, of gradient 1.0, corresponds to that of Figure 18.3

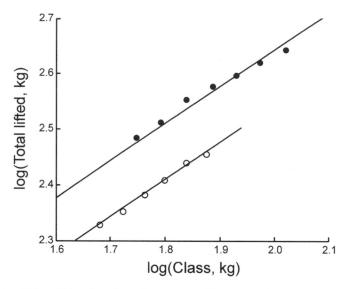

Figure 18.5 Weightlifting data from Chapter 14, with log (total lifted, kg) plotted against log (weight class, kg). The gradient of both lines is $2/3$

plotted on the x-axis and the logarithm of the total weight lifted that is plotted on the y-axis. The advantage does not relate here to the span of values, but to the interpretation of the data. In Section 12.5 it was argued that the total weight lifted should be proportional to (body mass)$^{2/3}$. Specifically,

for men,   (total lifted) $= 20.5 \times$ (body mass)$^{2/3}$

for women,   (total lifted) $= 16.1 \times$ (body mass)$^{2/3}$.

Taking logarithms of the both sides of each equation we get:

for men, $\log$ (total lifted) $= \log 20.5 + {}^{2}/_{3} \log$ (body mass)

$= 1.31 + {}^{2}/_{3} \log$ (body mass)

for women, $\log$ (total lifted) $= \log 16.1 + {}^{2}/_{3} \log$ (body mass)

$= 1.21 + {}^{2}/_{3} \log$ (body mass).

These are the equations of the two diagonal lines in Figure 18.5. Incidentally, if we were concerned with statistics, we might have started the analysis with this logarithmic plot, fitting lines to the data by linear regression analysis. This way we could have tested whether the resulting regression coefficients depart significantly from $2/3$.

18.5 What about units?

There are two unrelated points to be made under this heading. The first concerns what people speak of as 'log units'. Two quantities are said to differ by one log unit when one is 10 times the other. Thus 2 and 20 differ by one log unit because their logarithms are respectively 0.301 and 1.301, a difference of 1.000.

The second point has be addressed because of my emphasis on unit analysis, but it has little bearing on one's ability to think quantitatively. This point is that logarithms can strictly only be taken of numbers, and not of quantities involving physical units such as kilograms or litres. Most people go ahead regardless (including in scientific papers) and that causes no difficulties provided that the units are clearly specified. Often the units are specified within the mathematical expression itself as, for example, 'log (mass in kg)' or 'log (mass, kg)'. Other people get rid of the units before taking logarithms by dividing by a standard quantity with the same units. Thus, in the case of quantities measured in kilograms, each is divided by a standard 1 kg before logarithms are taken. This may be indicated by writing, say, 'log (mass/kg)'. It is unclear whether this trick accomplishes more than just appeasing the conscience.

A third possibility is to write, for example

$$\log (2 \text{ kg}) = 0.301 + \log (\text{kg}).$$

This avoids sweeping the problem under the carpet and allows the algebra of units to continue. However, I am unaware that anyone does this.

18.6 Natural logarithms

Logarithms to bases other than 10 are possible. For most purposes mathematicians favour logarithms that have e as their base, where $e = 2.718...$. These are known as natural, or Napierian, logarithms. For these, one may write $\log_e x$ or $\ln x$. Logarithms to base e and base 10 may be interconverted by means of the following formulae:

$$\log_{10} x = 0.4343 \times \log_e x. \tag{18.7}$$

$$\log_e x = 2.3026 \times \log_{10} x. \tag{18.8}$$

The '0.4343' in Eq. (18.7) is both $\log_{10} e$ and $1/\log_e 10$. The 2.3026 in Eq. (18.8) is both $\log_e 10$ and the reciprocal of 0.4343. Equations (18.3)–(18.6) apply to natural logarithms as well as to logarithms to the base 10.

Problems

18.1   See Section 18.4.

18.2   Simplify each of the following expressions, in each case combining three logarithmic terms into one. (a) $\log A - \log B + \log C$. (b) $\log A + \log B +$ D $\log E$.

18.3   What are the antilogarithms of the following? (a) $[\log A + \log B - \log C]$. (b) $3 \log D$.

18.4   Without using a calculator, give the logarithms of 20, 2000 and 0.02 to five places of decimal.

18.5   Without using a calculator, give the logarithms of 0.5 and 1/5.

18.6   Using values of logarithms given in the text, calculate (a) $\log (^5/_8)$ and (b) $\log (1.6 \times 4)$.

18.7   What is the logarithm of $10^{-0.43}$?

18.8   The value 7^2 almost equals 50, which is 100/2. From this, calculate an approximate value for $\log 7$.

18.9   Using $3^4 = 81$ and $80 = 2^3 \times 10$, calculate an approximate value for $\log 3$.

19

Exponential time courses

A quantity, call it y, is said to increase or decrease with time exponentially if it does so at a rate that, at any instant, is proportional to its current value. Another feature of exponential time courses is that the relative change is the same for any time interval of a given length. Both of these are true of compound interest when this is calculated continuously. Figure 19.1 illustrates an exponential rise and an exponential fall. People often misuse the phrase 'increases exponentially', using it just to describe a rapid increase.

A time course is exponential if the absolute value of a variable y changes at a rate that is proportional to its current value.

An exponential increase occurs, for example, with a population of bacteria when there are no deaths and the rate of division, and hence the rate of increase, is at all times proportional to the existing number of bacteria. Because the rate of increase itself increases with the size of the population, the slope of the curve rises with time, just as seen in Figure 19.1.

The general formula describing an exponential rise is:

$$y = y_0 e^{bt}, \tag{19.1}$$

where t is time and y_0 is the value of y when $t = 0$ (recall that $e^0 = 1$). y_0 cannot be zero, small as it looks in Figure 19.1. b is a positive constant, known

---

Biomedical Calculations   Richard F. Burton
© 2008 John Wiley & Sons, Ltd

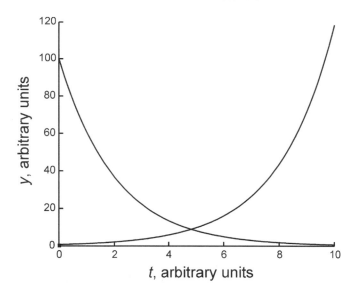

Figure 19.1 Examples of exponential time courses. The equations of the rising and falling curves are respectively $y = 0.8e^{0.5t}$ and $y = 100e^{-0.5t}$, where y and t have arbitrary units

as the rate constant. It may be replaced by $1/\tau$, in which case τ is called the time constant.

19.1   If t is measured in minutes, what must the units for b be?

In physiology the exponential time courses most often encountered involve decreases over time. Examples include the decay of radioactivity in isotopes and the gradual disappearance of some drugs from the blood. Numbers of bacteria in the blood, deliberately killed and injected for experimental purposes, may also fall exponentially as they are removed by the liver and spleen.

The general formula describing an exponential decrease with time could be taken as Eq. (19.1), but with b being negative. However, it is helpful and usual to make the negative nature of the rate constant explicit, with b still being taken as positive:

$$y = y_0 e^{-bt}. \tag{19.2}$$

When y is falling, the time course may be specified in terms of the 'half-time', commonly abbreviated as $t_{1/2}$. It is the time taken for y to fall to half some earlier value. This is the same wherever one starts on the curve. The half-time may be obtained directly from the curve by inspection, and its meaning is therefore more directly appreciated than that of b. The two are related:

$$t_{1/2} = (\ln 2)/b = 0.693/b. \tag{19.3}$$

'Half-time' is analogous to 'half-life' as applied to the exponential decay of radioactivity. For an exponential increase, one may likewise specify a 'doubling time'.

19.1 Use of semi-logarithmic plots

In Figure 19.1 a large part of each curve lies very close to the y-axis, so that the lowest y values are hard to compare as distinct data points. To achieve a more even spacing of y data, one might therefore choose to plot them as logarithms. As it happens, there is a better reason for doing so, as the curves become straight lines (Figure 19.2).

The two time courses can obviously be characterized by the gradients of these straight lines. In these particular examples they are 0.217 and -0.217. They equal $0.4343b$ and $-0.4343b$, respectively, where $0.4343 = \log_{10} e = 1/\log_e 10 = 1/2.303$. Indeed, in the general case, the gradient is 0.4343 times the rate constant. Alternatively, the rate constant is found as the gradient divided by 0.4343.

The last paragraph is largely implicit in Eq. (19.2). Taking logarithms of both sides and introducing brackets we have:

$$\log y = \log\{y_0(e^{-bt})\}.$$

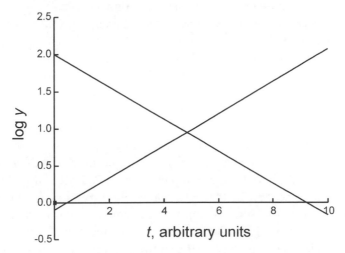

Figure 19.2   This is the same as Figure 19.1 except that $\log_{10} y$ is plotted instead of y

Applying Eq. (18.3), and then Eq. (18.5), we get:

$$\log y = \log y_0 + \log(e^{-bt}) = \log y_0 - bt \log e = \log y_0 - bt \times 0.4343$$
$$= \log y_0 - 0.4343bt.$$

This is the equation of a straight line. It tells us that $\log y = \log y_0$ at zero time ($t = 0$) and that the gradient is $-0.4343b$.

Figure 19.2 uses $\log_{10} y$ rather than $\log_e y$ because most biologists prefer the former. However, there is an advantage to plotting $\log_e y$, as then the gradients simply equal the rate constants.

19.2 Common complications

Time courses are not always as simple as have been described, but the above analysis is a good basis for understanding more complicated situations. Here, very briefly, are two examples.

When the temperature, T, of a non-living object rises or falls passively towards the temperature of the environment, T_e, it is not T that moves towards zero, but $(T - T_e)$. Equation (19.2) then becomes:

$$(T - T_e) = (T - T_e)_0 e^{-bt} = (T_0 - T_e)e^{-bt}. \tag{19.4}$$

This minor complication of the inclusion of a second term analogous to T_e is seen whenever the variable of interest changes exponentially with time towards some non-zero fixed value.

There have been many studies on the disappearance of substances injected into the blood. The usual complication then is that there is an initial period during which the substance mixes with the blood and also becomes diluted as some of it passes into the interstitial fluid (and maybe elsewhere). The resulting time course can often then be resolved into two more-or-less distinct exponential phases. From a linear semi-logarithmic plot of the second phase one may then be able to extrapolate back to time zero (the moment of injection) to estimate what the concentration would have been if complete mixing had been instantaneous. It was pointed out in Section 5.4 that extracellular fluid volumes may be estimated using marker substances such as inulin, as the amount injected divided by the concentration after complete mixing. It is by extrapolating back in time that allowance may be made for losses of the substance during the mixing period.

Problems

19.1 See the beginning of this chapter.

19.2 After injection or infusion into the blood inulin and iothalamate are each re-moved by the kidneys at rates proportional to the product of the glomerular filtration rate and the concentration in the blood plasma. What conditions would have to be satisfied for the plasma concentration to fall exactly exponentially?

19.3 Here are some values of y that fall exponentially with time t.

t (min)	0	1	2	3	4	5	6
y	100	70	49	34.3	24	16.8	11.8

(a) Plot y against t.

(b) Over the first minute y falls by a factor of $70/100 = 0.7$. Verify that y falls by the same factor over all the other 1 minute periods.

(c) You can see that y falls to half its original value of 100 in just under 2 min — actually 1.94 min. This period of 1.94 min is $t_{1/2}$. Verify that y is slightly more than halved over any period of 2 min.

(d) Now calculate $\log_{10} y$ for each y and plot $\log_{10} y$ against t. Do you get a straight line? What is the gradient? What is the rate constant?

(e) Do you also get a straight line using natural logarithms?

(f) It has been emphasized that, when y falls exponentially, it changes at a rate that is proportional to its instantaneous value. Check that this is so here, i.e. that the gradient of the curve is always proportional to y. Three of the gradients are given for you below (as the change in y per minute). Estimate (approximately) the missing one from the graph. Do this by drawing a tangent to the curve at $t = 2$ min (the time at which $y = 49$).

y	100	—	49	—	—	16.8	11.8
gradient (min^{-1})	-35.6	—	?	—	—	-6.0	-4.2

Plot these gradients against y to see if the points fall on a line through the origin. Such a line implies a constant ratio of gradient to y.

(g) Calculate this ratio. How does it relate to the gradient calculated in (d)?

(h) Write out Equation (19.2), but using numerical values for y_0 and b.

20

Nernst equations in physiology and biochemistry: logarithms and 'RT/zF'

What is *the* Nernst equation? Physiology textbooks, usually in their early chapters, give an equation by that name that relates equilibrium potentials to ionic concentrations, usually in the context of nerve or muscle membranes. Biochemistry textbooks apply the name to a somewhat similar-looking equation relating redox potentials to concentrations of oxidants (electron acceptors) and reductants (electron donors) in oxidation–reduction reactions. According to Walter Nernst's own textbook, the two equations are related, but neither is *the* Nernst equation. My aim here is to focus on features shared by both, but concentrating mainly on the physiologists' Nernst equation. The other involves concepts – redox potentials and standard potentials – that require more description and are better obtained from biochemistry textbooks.

The physiologists' Nernst equation describes the relationship between the concentrations of a particular inorganic ion either side of the cell membrane and the associated electrical potential difference that corresponds to diffusion equilibrium across the membrane. This is called the 'equilibrium potential' for that particular ionic species, but there is no implication that an equilibrium actually exists. To avoid mistakes over + and − signs in the mathematical treatment, it helps to start by recalling how the polarities of membrane potentials are defined. By convention, the resting membrane potential of a typical cell is described as negative, with the

Biomedical Calculations   Richard F. Burton
© 2008 John Wiley & Sons, Ltd

inside being negative with respect to the outside. The equilibrium potentials of potassium and chloride ions are also negative, typically (for chloride especially) being somewhat similar in magnitude to the resting potential. The equilibrium potential for sodium has the opposite polarity.

To consider the case of an ionic species X of valency z, if the internal and external concentrations are $[X]_i$ and $[X]_e$ respectively, then one form of the physiologist's Nernst equation is as follows:

$$E_x = \frac{RT}{zF} \ln \frac{[X]_e}{[X]_i}. \tag{20.1}$$

E_x is the equilibrium potential. T is the absolute temperature. The constant R is the same gas constant as is discussed in Chapter 17, although here it has nothing to do with gases. F is a constant called the Faraday, a measure of ionic charge ('specific ionic charge'), which equals 9.65×10^4 J volt^{-1}equiv^{-1} or 2.306 $\times 10^4$cal volt^{-1}equiv^{-1}. z is the valency of X, so that there are z equivalents of a given ion in a solution for each mole (Section 2.7). Ideally, ion activities should be used in preference to concentrations. (Readers who know more about electricity than is the concern of this book may perceive that the Faraday also equals 9.65×10^4 coulombs per equivalent, as 1 coulomb is 1 J volt^{-1}.)

20.1   To check that the ratio $[X]_e/[X]_i$ is the right way up, try substituting realistic values for potassium, say 5 and 150 mmol/kg water respectively for the external and internal solutions. Calculate $\ln [X]_e/[X]_i$. Is its sign negative, as it should be?

20.2   Equation 20.1 is sometimes written with $[X]_e$ and $[X]_i$ interchanged. What other change is then needed in the equation?

20.1 More on RT/zF

The expression $[X]_e/[X]_i$ is free of units and so too is its logarithm. This means that RT/zF has the same units as E_x, namely V or mV (the standard symbols for volt and millivolt respectively).

The units used for R and F must be compatible with each other. One version of R is 1.987 cal mol^{-1} K^{-1} (Section 17.4) and that is appropriately paired with the above 'calorie' version of F, i.e. 2.306×10^4cal V^{-1}equiv^{-1}. We may now find the value of RT/zF in mV for positive univalent ions ($z = 1$) at

37 °C (310 K):

$$\frac{RT}{zF} = \frac{1.987 \text{ cal}}{\text{mol K}} \times 310 \text{ K} \times \frac{1 \text{ mol}}{1 \text{ equiv}} \times \frac{\text{V equiv}}{2.306 \times 10^4 \text{ cal}} \times \frac{1000 \text{ mV}}{1 \text{ V}} = 26.7 \text{ mV}.$$

In this case, therefore,

$$E_x = 26.7 \ln \frac{[X]_e}{[X]_i} \text{ mV}. \tag{20.2}$$

Given the common preference for common logarithms, Eq. (20.1) may be altered accordingly. Since $\ln a$ (or $\log_e a$) $= 2.303 \log a$,

$$E_x = 2.303 \frac{RT}{zF} \log \frac{[X]_e}{[X]_i}. \tag{20.3}$$

Therefore, for positive univalent ions at 37 °C:

$$E_x = 61.5 \log \frac{[X]_e}{[X]_i} \text{ mV}. \tag{20.4}$$

20.3   Calculate E_x (i.e. E_K) at 37 °C for the potassium concentrations used above, namely 150 and 5 mmol/kg water, respectively, for the internal and external solutions.

The equilibrium potential thus calculated is more negative than the resting membrane potentials of most cell types. Potassium is therefore not in diffusion equilibrium; its tendency to diffuse out is opposed by the sodium–potassium pump.

I now present the 'other' Nernst equation here merely to illustrate points of similarity with the above version. Here it is in a form analogous to Eq. (20.1):

$$E - E_0 = \frac{RT}{nF} \ln \frac{[\text{oxidant}]}{[\text{reductant}]}. \tag{20.5}$$

E is the redox potential and E_0 is its standard value. Taking the place of z is n, the number of electrons involved per molecule. Much of the above discussion is clearly applicable here.

It is relevant to both physiology and biochemistry that Eq. (20.1) applies also to ion-selective electrodes, including pH electrodes. In the latter case it is a membrane of special glass that separates the two solutions, the inner one of which has a constant known pH. If we apply Eq. (20.3), taking $[H^+]_i$ as constant and simplifying, we get:

$$E_H = A \log [H^+]_e + B, \tag{20.6}$$

where A and B are temperature-dependent constants. If we take pH as $-\log [H^+]$, then the voltage across the electrode glass should be proportional to pH (which it is). Unfortunately, the situation is not so simple, since actually one must consider activities of hydrogen ions and not concentrations. There are also practical and conceptual problems associated with the necessary use of reference electrodes to complete the electrical circuit. In short, as discussed in the next chapter, the true definition of pH is not really $-\log [H^+]$.

Problems

20.1 See the beginning of this chapter.

20.2 See the beginning of this chapter.

20.3 See Section 20.1.

20.4 If the extracellular and intracellular concentrations of chloride are respectively 110 and 5 mmol/kg water for a particular cell, what is the equilibrium potential for chloride at 37 °C?

20.5 If the extracellular and intracellular concentrations of bicarbonate are respectively 26 and 13 mmol/kg water, what is the equilibrium potential for bicarbonate at 37 °C? What can you conclude about the distribution of bicarbonate across the cell membrane?

20.6 Consider a non-biological membrane separating two solutions at 37 °C, one with a calcium concentration 10 times that of the other. Reasoning from Eq. (20.4) for univalent ions, calculate the equilibrium potential for calcium (ignoring the matter of polarity).

21
pH – two definitions and a possible dilemma for teachers

According to the definition of pH given by most textbooks of general chemistry, biochemistry and physiology,

$$pH \equiv -\log[H^+], \qquad (21.1)$$

where $[H^+]$ stands for the molar concentration of hydrogen ions, or more correctly hydronium ions, 'H_3O^+', for which 'H^+' may be taken as shorthand. Although this looks completely straightforward, this definition was rejected long ago by physical chemists and it is not the one underlying routine measurements of pH. The theoretical basis of the modern definition is unsuited to elementary teaching, but in simple terms pH may be defined as a reading on a properly maintained and calibrated pH meter. Put another way, this 'conventional' pH scale is defined by a set of standard buffer solutions that have each been selected, on rigorous theoretical grounds, as representing particular pH values.

An advantage of the old definition is that $[H^+]$ and $[H_3O^+]$ are themselves straightforward concepts, but it has two disadvantages: firstly that it is unsatisfactory and secondly that it is frequently taught to students who have little understanding of logarithms. An advantage of the newer, conventional scale is that it can be applied in many contexts without any reference to logarithms (but not all, as is illustrated in the next two chapters). The main disadvantages of the newer scale are those of swimming against the tide and, for most people, of

Biomedical Calculations Richard F. Burton
© 2008 John Wiley & Sons, Ltd

having to take its theoretical justification on trust. For teachers and textbook writers aiming at simplicity there could thus be a dilemma here, but the solution is probably to combine both approaches in whatever way seems most appropriate to the supposed future needs of particular students. There is, in any case, no harm in revealing from time to time that science can be more complicated and interesting than elementary textbooks reveal. The scales are discussed in more detail below, but first some general points must be made that apply to both. The Notes provide further details.

pH values generally range between zero and 14, although these are neither strict limits, nor at all closely approached in living animals. They are free of units in the usual sense (e.g. units of concentration) as is necessarily true of logarithms. However, regardless of that, it is often helpful to speak of 'pH units', with one pH unit being the difference between one whole-number pH value and the next. Confusing to some people initially is the fact the numbers low on the pH scale correspond to high acidity, while high numbers correspond to alkalinity (low concentrations of hydrogen ions). This, with either definition, is an especially important point to be clear about.

21.1 pH as $-\log[H^+]$

The definition of pH in terms of Eq. (21.1) and the molar concentration of hydrogen (or hydronium) ions was proposed by Sørensen in 1909. It is the negative sign that makes acid solutions have low pH values. The units of concentration are dropped in the process of taking logarithms (see Section 18.5).

Equation (21.1) has an alternative, non-logarithmic, form that allows the hydrogen ion concentration (in mol/L) to be calculated from the pH:

$$[H^+] = 10^{-pH}. \tag{21.2}$$

As an example, we may start with pH 7, which is approximately the pH of neutrality and a common pH inside cells. From Eq. 21.2, the hydrogen ion concentration is 10^{-7} mol/L, which one may choose to express as 100 nmol/L. Note first that this is an extremely low concentration; there is therefore no possibility of measuring it by titration, nor of preparing a solution of that pH just by measuring out quantities of strong acid or alkali. Another point to note is that a low concentration certainly does not indicate biological unimportance; enzymes and cell mechanisms generally are very sensitive to changes in pH near neutrality.

Transmembrane proton gradients are fundamental to the functioning of mitochondria and it is particularly revealing to calculate the concentration of hydrogen ions inside the matrix of a small one and to consider its implications. There is no definitive pH to be quoted here, but we may take pH 8 as a round number value. According to Eq. (21.2), the concentration is then 10^{-8} mol/L. A representative value for the volume of a small mitochondrion was calculated in Chapter 11.1 as 0.5 μm^3. What, then, is the approximate *number* of hydrogen ions in the matrix of such a mitochondrion? Recall that 1 mol of a substance corresponds to 6.0×10^{23} molecules or, in this case, hydrogen ions (this being Avogadro's number – Section 2.6). To start the calculation we may convert 0.5 μm^3 to litres: there are $10^{15} \mu m^3$ per litre, so that the mitochondrial volume is 0.5×10^{-15} L. The rest of the calculation is as follows:

$$0.5 \times 10^{-15} \text{ L} \times 10^{-8} \text{ mol/L} \times 6.0 \times 10^{23} \text{ ions/mol} = 3 \text{ ions.}$$

This does not allow for the intermembrane space. Moreover, some mitochondria have less than a third of the chosen volume. A number less than 1 is surely food for thought! Should we regard it as an average over time?

Here is another disquieting fact in relation to Eq. (21.1) as a definition of pH. The pH of 10^{-1} M hydrochloric acid should be 1.0. If its pH is measured by means of a pH meter that has been accurately calibrated against standard buffer solutions, the actual reading is 1.088 (at 25 °C).

---

pH is a number on a conventional scale. It does not relate exactly to the hydrogen ion concentration.

21.2 The true definition of pH: pH as a number on a conventional scale

Difficulties with the above definition of pH were recognized by Sørensen just a few years after he proposed it. Actual concentrations of hydrogen ions can only be measured directly when they are quite high and the responses of H^+-sensitive electrodes, including the familiar glass electrodes, relate not to concentration, but to something related, that is called 'activity' (Section 9.10). Even 'activity' is not quite accurate, because the activity of a single ionic species, as opposed to that of anion-cation pairs, is not strictly definable thermodynamically. As a non-technical term, 'effective concentration' conveys something of the right idea.

A related problem is that pH-sensitive glass electrodes must always be used in conjunction with a second (reference) electrode to complete the circuit and this too is somewhat affected by the solution under test. The individual contributions of the two electrodes to the electrical potential difference between them cannot, in principle, be exactly assessed. Each has thus been likened to the sound of one hand clapping.

Because of these difficulties, the pH scale now in most general use has to be a conventional one, defined operationally in terms of standard buffer solutions. Thus, as an example of various primary standards, a solution containing 0.05 mol kg^{-1} potassium hydrogen phthalate is defined as having a pH of 4.000 at 20 °C. Simplistically stated, pH is then defined as 'the reading on a properly calibrated pH meter'. There is of course much more to the matter than that, and the scale is far from arbitrary (see Notes). However, for most biologists, the exact compositions and rationale of the standard pH buffer solutions are unimportant and it is rarely necessary to consider pH in terms of solution theory. It is usually enough to accept the numbers as a practical scale of acidity.

For a pH value to have practical meaning, it is usually compared with some other number on the same scale. Clinicians can learn much from measurements of arterial blood pH without necessarily understanding the pH scale; it is the relationship of measured values to the normal value of 7.4 that counts, or else to other values that are known to be life-threatening. *In vitro* media can be adjusted to empirically known optimum values that lack any theoretical basis. For gardeners, it suffices to know that a soil pH of 5 is satisfactory for rhododendrons without having to dig deeper into pH theory. All this is accomplished just by comparing numbers – actual pH readings compared with norms or optima.

pH mainly matters within the body because it affects the properties of proteins and other molecules. In regard to proteins it does so through the ionization of individual amino, carboxyl and especially imidazole groups. We will come back to that in relation to buffers (Chapter 23), but the point to be made just now is that the ionization state of a particular substance or protein group is determined by the difference between pH and another number, called the pK, that is characteristic of the substance or protein group in question. The numerical value of that pK is measured on the pH scale, so the difference is independent of how the pH scale is defined. The pK value is the pH at which the substance is half ionized and half unionized.

21.3 The meaning of 10^{-pH}

It is often useful to use a measure of hydrogen ions that is more like the actual concentration, $[H^+]$, in form. By this I mean the quantity calculated as 10^{-pH}.

For this, the symbol adopted here is $\{H^+\}$. In other words,

$$\{H^+\} \equiv 10^{-pH}. \tag{21.3}$$

This resembles Eq. (21.2), while Eq. (21.1) is replace by

$$pH = -\log\{H^+\}. \tag{21.4}$$

In terms of the first definition of pH and Eq. (21.2), $\{H^+\}$ takes the place of the molar hydrogen ion concentration. According to the conventional scale and the usual calibration procedures, it corresponds more strictly to the activity of hydrogen ions inasmuch as that is definable (see above). It is unfortunate that there is no completely satisfactory one-word name for what is actually then calculated as 10^{-pH}, because nameless entities tend to be ignored or misunderstood. The word 'activity' is often used, but 'conventional activity' or 'notional activity' would be more correct. A special unit would be useful too if it existed (the 'Sorensen' perhaps?), and that would obviate the need to specify litres of solution or kilograms of water. Special symbols, like $\{H^+\}$, are not a problem, because all symbols should be defined whenever they are used.

21.4 Final comments

Note that pK and pH are defined in relation to the same scale. It is the difference between them that determines the ionization and associated properties of an acid or base. In general pH has little significance as an isolated number, but only in comparison with other numbers on the same scale. From that limited point of view, it would make no difference if the whole pH scale had been defined as if shifted up or down by some constant amount.

To focus on the concept of pH, as here, is not to deny the importance of hydrogen ion *amounts* and *concentrations* – in relation, for example, to the stoichiometry of reactions, titrations of acid or alkali, and the acid content of stomach acid. What should be clear now is that pH measurements do not provide exact measures of hydrogen ion concentration through Eq. (21.2). Nevertheless, one should not so wholeheartedly reject '$[H^+]$' in favour of '$\{H^+\}$' as to be unaware either of their similarity or of their minuteness near neutral pH.

Problems

21.1 A sample of gastric juice has a pH of 1.1. What is the *approximate* hydrogen ion concentration?

21.2 Some medical textbooks express the acid–base status of arterial plasma in terms, not of pH, but of hydrogen ion concentrations (in nmol/L) calculated from pH in accordance with Eq. (21.2) (i.e. treating $\{H^+\}$ as $[H^+]$). (a) Express the normal arterial pH of 7.4 this way. (b) Calculate the equivalent concentration for pH 7.1. (For an instant answer, make use of the value of log 2 emphasized in Section 18.3.) (c) Would these concentrations be exactly correct?

21.3 The normal range of arterial blood pH is often given as 7.35–7.45. Treating $\{H^+\}$ as $[H^+]$, convert these two values to concentrations in nmol L^{-1}, to two significant figures only. Might the coincidence of numbers help as a memory aid?

22

Equilibrium constants, the Henderson–Hasselbalch equation, dose–response curves

Much of biochemistry and physiology has to do with reversible chemical reactions in which two substances, call them A and B, combine to form a compound AB which can then dissociate to some extent back to A and B. Think
of oxygen binding to myoglobin, calcium to citrate or proteins, hydrogen ions
to buffers, substrates to enzymes and drugs to receptors. Our concern here is
with the equilibrium condition in which the concentrations of the reactants are
stable – meaning that A and B are combining at the same rate as AB is disso-
ciating. Logarithms are involved again in relation to the associated algebra and
graphs. The topics include dissociation constants, the logarithmic Henderson–
Hasselbalch equation, Michaelis–Menten kinetics and semi-logarithmic dose–
response curves. The emphasis is more on a unity of approach than on specific
reactions, and more on fostering a basic understanding than on treating more complicated situations. Some of these ideas are developed in the next chapter in relation
to buffering.

Biomedical Calculations   Richard F. Burton
© 2008 John Wiley & Sons, Ltd

22.1 Equilibrium constants

Consider a situation in which our two substances, A and B, combine to form a third one, AB, according to the following reversible reaction:

$$A + B \leftrightarrows AB.$$

With square brackets representing concentrations, the relationship amongst the reactants at equilibrium is given by:

$$\frac{[A][B]}{[AB]} = K, \qquad (22.1)$$

where K is called the equilibrium constant. It is usual, though not necessary, to have the dissociation products on the top line (i.e. A and B), in which case K may be called the dissociation constant of AB. If A or B is ionic, K may also be called an ionization constant. K may also be defined as $[AB]/([A][B])$, in which case it is the 'association constant'. Numerical values depend on the particular reaction, but also on physical conditions such as temperature.

> For the reaction $A + B \leftrightarrows AB$,
>
> dissociation constant $K = \dfrac{[A][B]}{[AB]}$

Equation (22.1) is easily derived. According to the law of mass action, the rate at which A and B combine is proportional to the concentration of each. Likewise, the rate at which AB dissociates is proportional to the concentration of AB. With the constants of proportionality being respectively k_1 and k_2:

$$\text{rate of association} = k_1[A][B],$$

$$\text{rate of dissociation} = k_2[AB].$$

At equilibrium, these rates are equal, by definition. Therefore:

$$k_1[A][B] = k_2[AB]$$

and

$$\frac{[A][B]}{[AB]} = \frac{k_2}{k_1}.$$

It just remains to replace the ratio k_1/k_2 with K.

One of the reactants may be water, as in the formation of carbonic acid:

$$CO_2 + H_2O \leftrightharpoons H_2CO_3.$$

In such cases the concentration of water is conventionally taken as 1, so that '$[H_2O]$' does not appear in the expression for the equilibrium constant. Thus,

$$K = \frac{[CO_2]}{[H_2CO_3]} \tag{22.2}$$

If $[H_2O]$ were included, it would be on the top line. Note in passing that the actual concentration of water at $4\,^\circ C$ is 55.56 M, calculated as 1000 g/L $\div$ 18 g/mol.

The right-hand side of Eq. (22.2) is perhaps more often written the other way up, i.e. as its reciprocal, the association constant, but either version is valid. For consistency with Eqs (22.1) and (22.3), I have made K the dissociation constant.

If a reaction proceeds in stages, i.e. as a chain of separate reactions, an equilibrium constant may be defined for the chain as a whole. Because carbonic acid, formed as above, dissociates to produce bicarbonate and hydrogen ions, one may treat the following overall reaction without reference to carbonic acid:

$$CO_2 + H_2O \leftrightharpoons HCO_3^- + H^+.$$

Then

$$K = \frac{[HCO_3^-][H^+]}{[CO_2]}. \tag{22.3}$$

An overall reaction may involve more substances, and perhaps more than one molecule of each. For example, it may be that a molecules of A combine with b molecules of B to form the complex A_aB_b. In that case:

$$K = \frac{[A]^a[B]^b}{[A_aB_b]}.$$

In a more complicated situation, the reaction might take the form:

$$aA + bB \leftrightharpoons cC + dD + eE,$$

so that:

$$K = \frac{[A]^a [B]^b}{[C]^c [D]^d [E]^e}.$$

22.2 Concentrations or activities?

So far the equilibrium constants have been defined in terms of concentrations, but activities are often used instead (Section 9.10). Indeed, it is impractical to use the hydrogen ion concentration, $[H^+]$, rather than $\{H^+\}$ or 10^{-pH} and $\{H^+\}$ should therefore replace $[H^+]$ in Eq. (22.3). Concentrations are otherwise what are generally measured and are therefore of most practical significance. The use of activities gives equilibrium constants that are less affected by the concentrations of ions that are not themselves involved in the reactions. When based on concentrations, K may be called an 'apparent equilibrium constant' and designated K'.

22.3 The Henderson–Hasselbalch equation

This equation relates specifically to equilibria involving hydrogen ions (protons), acids (proton donors) and bases (proton acceptors). These can be represented thus:

$$C^- + H^+ \leftrightharpoons CH$$

and

$$C + H^+ \leftrightharpoons CH^+.$$

C and C^- are bases. CH and CH^+ are acids. The dissociation constant is given by:

$$K = \frac{[base]\{H^+\}}{[acid]}.$$

Karl Hasselbalch re-expressed this in logarithmic terms:

$$\log K = \log \frac{[base]\{H^+\}}{[acid]} = \log \left(\frac{[base]}{[acid]} \right) + \log \{H^+\}.$$

Rearranged, this gives:

$$- \log \{H^+\} = \log \left(\frac{[\text{base}]}{[\text{acid}]} \right) - \log K.$$

Just as $-\log \{H^+\}$ should now be replaced with pH, so, by analogy, '$-\log K$' may be written as 'pK'. Then:

$$pH = pK + \log \left(\frac{[\text{base}]}{[\text{acid}]} \right). \tag{22.4}$$

This is the Henderson–Hasselbalch equation. We meet this below, adapted to the special case of the $CO_2/HCO_3^-/H^+$ equilibrium.

Rearrangement of Eq. (22.4) gives:

$$\log \left(\frac{[\text{base}]}{[\text{acid}]} \right) = pH - pK,$$

or

$$\frac{[\text{base}]}{[\text{acid}]} = 10^{(pH - pK)}. \tag{22.5}$$

The problem of defining pH is discussed in Chapter 21, but the conceptual and practical difficulties vanish when one considers the difference, $(pH - pK)$. This is because pK values are determined in relation to the pH scale.

Here is an opportunity to think about units again. The expression [base]/[acid] is the ratio of two quantities (i.e. concentrations) that have the same units. Therefore the ratio is free of units too, and it does not matter whether, for example, the individual concentrations are molar or molal, as long as they are the same. There is no problem with taking the logarithm of this unit-free quantity, and the logarithm, like all logarithms, is free of units too.

22.4 Application of the Henderson–Hasselbalch equation to drugs

At least 60 per cent of drugs undergo acid–base ionization reactions in aqueous solution. Whether or not such drugs are ionized (i.e. carry positive or negative charges) has an important influence on their solubility, activity and ability to pass through cell membranes. As a general rule, they permeate these more readily when they are uncharged, and are also more easily absorbed from the gut. Equation

(22.5) can be used to calculate the ratio of base to acid forms of a drug from its pK and the pH. For example, phenobarbital has a pK of about 7.4 and the ratio at pH 7.4 (as in arterial plasma) is therefore about 1.0; the two forms are therefore present in approximately equal amounts. Recall that an acid is a proton (hydrogen ion) donor and a base is a proton acceptor.

In this context one must distinguish two categories of ionizing drug. In some the acid form is uncharged (XH) and the basic form is charged (X$^-$). Drugs with carboxyl groups are in this category. In other drugs the acid form is charged (XH$^+$) and the basic form is uncharged (X), as applies to most amines. Either way, the basic form is always favoured at high pH. Drugs in the first category (with their approximate pK values in brackets*) include furosemide (3.9), phenobarbital (7.4) and phenytoin (8.2). Drugs in the second category, with basic form uncharged, include caffeine (0.7), cimetidine (6.8), morphine (8.0) and erythromycin (8.8).

22.1   (a) Aspirin has a pK of 3.5 and has a negatively charged carboxyl group at high pH. What is the ratio of base to acid forms at pH 7.4? (b) Diazepam has a pK of 3.3 and a positive charge at low pH. What is the ratio of base to acid forms at pH 7.4? (c) If molecular charge were the only relevant factor, which one should pass through cell membranes most easily at pH 7.4?

Drugs are often categorized as 'acidic' or 'basic', irrespective of pH. What can this mean, when a given drug substance may exist in both acid and base forms? According to the commonest usage, a drug is called an acidic drug if the acid form is uncharged (XH) and the basic form is anionic (X$^-$). It is called a basic drug if it is the basic form that is uncharged (X) and the acid form is cationic (XH$^+$). According to another usage, less common and less helpful, acidic drugs have pK values below 7.4 and basic drugs have pK values above 7.4. Depending on how one defines 'acidic' and 'basic', several of the above drugs can be placed in either category! It may be to avoid misunderstandings that many textbooks avoid the terms 'basic drug' and 'acidic drug'. What matters are the pK and the type of ionization.

22.5 The dependence of [AB] on [A] when ([B] + [AB]) is constant

A situation often to be considered is one in which the concentration of one reactant varies, say A, while the total concentration of the other in its two forms

---

* Different sources give slightly different values, determined perhaps under different conditions.

B and AB remains constant. Particular significance commonly then attaches to the concentration of [AB] and how this is affected by [A] when ([B] + [AB]) is constant. For example, one might be interested in how the concentration of oxygenated myoglobin (and thus of bound oxygen) varies with the concentration of dissolved oxygen. As another example, A could be ligand that binds to a receptor that exists as B and AB. A little algebra gives us the relationship for A and B. From Eq. (22.1),

$$[B] = K[AB]/[A].$$

Therefore,

$$([B] + [AB]) = K[AB]/[A] + [AB]$$

$$= [AB]\{K/[A] + 1\}.$$

Calling ([B] + [AB]) '(Total B)' and rearranging, we then have:

$$[AB] = \frac{(\text{Total B})}{K/[A] + 1} = \frac{(\text{Total B})}{(K + [A])/[A]}$$

$$= (\text{Total B}) \frac{[A]}{K + [A]}. \tag{22.6}$$

22.2   If [AB], (Total B) and [A] happen to be expressed in mmol/L, what units, if any, must apply to K, as judged entirely from the form of the equation?

Let us look more closely at the relationship between [AB] and [A] when ([B] + [AB]) is constant. Just from Eq. (22.6), one can work out three features of the relationship:

1. If [A] is negligibly small compared with K, the equation reduces to:

$$[AB] = (\text{Total B}) \frac{[A]}{K}.$$

   Then [AB] is virtually proportional to [A].

2. If [A] is very large compared with K, $[A]/(K + [A])$ approximates to 1 and [AB] is virtually independent of [A]. In other words, as [A] increases from zero, [AB] eventually rises to a plateau value that equals (Total B).

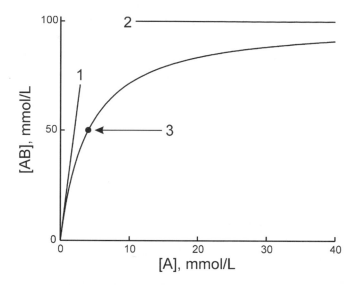

Figure 22.1   The relationship between [AB] and [A] according to Eq. (22.2). The numbered features are as explained in the text. Line 1 represents the gradient at low concentrations. Line 2 shows the upper limiting value (plateau) of [AB] and point 3 corresponds to [A] = K = 4 mmol/L. (Total B), i.e. ([B] + [AB]), is 100 mmol/L

3. When [A] equals K, $[A]/(K + [A])$ equals 0.5 and [AB] equals (Total B)/2. Figure 22.1 shows these three features, numbered, and also the actual curve for an arbitrary example where (Total B) is 100 mmol/L and [A] is 4 mmol/L. The curve is known as a rectangular hyperbola. Its formula may be generalized as follows:

$$Y = (\text{plateau value of } Y)\ \frac{X}{k + X}. \tag{22.7}$$

In this general form X and Y can have a variety of meanings, with k being some appropriate constant. In form it matches the curve in Figure 3.1 relating percentage body fat to the ratio of body fat to fat-free mass. More generally in biology it has been applied, for example, to carbon dioxide fixation by aquatic algae and to rates of feeding by insects and herbivores. In relation to enzyme kinetics it has the form of the *Michaelis–Menten equation*, with Y being the initial rate of an enzyme reaction (V), X being the concentration of substrate ([S]) and k being the Michaelis constant, K_M. With V_{max} being the plateau (i.e. maximum) value of V,

$$V = \frac{V_{max}[S]}{K_M + [S]}. \tag{22.8}$$

The derivation of the Michaelis–Menten equation is to be found in many biochemistry textbooks. It resembles the derivation of Eq. (22.1), but with an additional step, the conversion of substrate to final product.

22.3   What is the substrate concentration, [S], at which V is half-maximal?

22.6 Concentration–response curves and dose–response curves

Equation (22.7) commonly describes the relationship between the dose or concentration of a drug or hormone (X) and the response (Y) it produces *in vivo* or *in vitro*. This is true provided that the response is proportional to the amount of drug that is bound to the relevant receptors. Figure 22.2 shows, for three hypothetical drugs, the relationship between the response, as a percentage of maximum, and the dose of drug (amount or concentration) producing the response. As illustrated, the doses that are needed to produce half the maximum response vary 100-fold, but drugs may vary in potency much more than that. Partly for this reason, doses are very often plotted on a logarithmic scale, as is illustrated for the

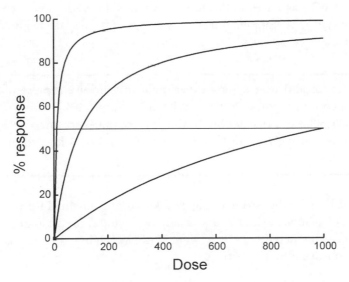

Figure 22.2   The relationship between percentage response and dose (in arbitrary units) for three hypothetical drugs

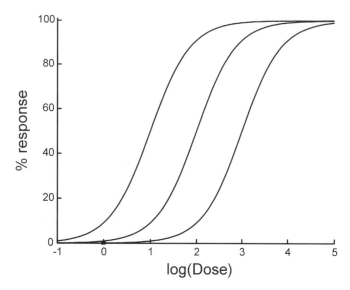

Figure 22.3 This shows the three relationships of Figure 22.2 with doses plotted on a logarithmic scale

same three drugs in Figure 22.3. The three curves share the same sigmoid shape, but are positioned differently along the dose axis. This is a convenient way of plotting dose–response curves even when they do not conform to Eq. (22.7). Plotted responses may be as varied as the accumulation of an intracellular second messenger, changes in membrane potential, secretion rate, heart rate and muscular contraction.

22.4   A typical dose–response experiment may involve the use of 10–20 doses or concentrations. They might be, for example, 0.3, 1, 3, 10, 30, 100, 300, 1000, 3000 and 10 000 nM. Although these values are unevenly spaced, they are chosen to give almost exactly equal spacing on the logarithmic scale. Explain this.

The equation for the curves in Figure 22.3 may be obtained from Eq. (22.7), taking the plateau value as 100 and changing X and k in accordance with the logarithmic scale. Recall that $10^{\log a} = a$. With X standing for the dose, we can therefore substitute $10^{\log(\text{dose})}$ for X:

$$\% \text{ response} = 100\frac{10^{\log(\text{dose})}}{10^{\log k} + 10^{\log(\text{dose})}}. \tag{22.9}$$

The gradient of the curve at its mid-point, where the response is 50 per cent of maximum, is $100/4 \times \ln 10 = 25 \times 2.303 = 57.56$.

Problems

22.1   See Section 22.4.

22.2   See Section 22.5.

22.3   See Section 22.5.

22.4   See Section 22.6.

$$\boxed{\frac{[\text{base form}]}{[\text{acid form}]} = 10^{(\text{pH}-\text{p}K)}}$$

22.5   People speak of lactic acid in the blood, but what is the ratio of lactic acid to lactate? Its $\text{p}K$ in plasma is about 4.6. Take the pH as 7.4. Their combined concentration in the plasma of a resting person is typically about 0.7–2.0 mmol/L.

22.6   Morphine has a positive charge at low pH and a $\text{p}K$ of 8.0. What percentage would be charged (protonated) at pH 7.0?

22.7   Local anaesthetics are more effective in their unionized (base) form and therefore at high pH. The $\text{p}K$ values of bupivacaine and lignocaine are respectively 8.1 and 7.9. Is this small difference significant? Calculate first the ratios of base to acid forms at pH 7.4 and then the percentages in base form.

22.8   The total concentration of inorganic phosphate in blood plasma is about 1–1.5 mM. Taking this total as 1.2 mM, what are the millimolar concentrations of the two forms, HPO_4^{2-} and $H_2PO_4^{-}$, at pH 7.4? The $\text{p}K$ is 6.8.

22.9   Although inorganic phosphate is usually treated in terms of the two forms $H_2PO_4^{-}$ and HPO_4^{2-}, there are two other equilibria:

$$PO_4^{3-} + H^+ \leftrightharpoons HPO_4^{2-}$$

$$H_2PO_4^{-} + H^+ \leftrightharpoons H_3PO_4$$

The $\text{p}K$ values are respectively 12.4 and 2.0 at body temperature. At pH 7.4, what are the ratios of $[PO_4^{3-}]$ to $[HPO_4^{2-}]$ and of $[H_2PO_4^{-}]$ to $[H_3PO_4]$?

22.10   One may choose to express the dissociation of acetic acid in either of the two following ways:

$$CH_3COOH \leftrightharpoons CH_3COO^- + H^+$$

or

$$CH_3COOH + H_2O \leftrightharpoons CH_3COO^- + H_3O^+.$$

Why is the dissociation constant unaffected by the choice?

22.11   The following calculations illustrate a method of estimating intra-cellular pH that involves the weak acid DMO (5,5-dimethyl-2.4-oxazolidinediione) (see Notes). This has the property of diffusing through cell membranes only in the uncharged acid form (like some drugs). After cells have been allowed to equilibrate with a solution containing DMO, the concentration of the acid form inside them equals that outside. How much of the base is produced by dissociation of the acid inside the cells depends on the intracellular pH. The latter is calculated from the pK ($= 6.13$ at 37 °C), the extracellular pH and the total extracellular and intracellular concentrations. There are four stages in the calculation of the intracellular pH, but each is no more complicated than those already tackled.
(a) The extracellular pH is 7.40. Calculate the ratio [base]/[acid] in the extracellular solution. (b) The extracellular concentration, (acid + base), is 1 mmol/kg water. Calculate the concentration of DMO in the permeant (acid) form. After equilibration this is the same inside and out. (c) If the total intracellular concentration of DMO in both forms is found to be 0.35 mmol/kg water, what is the concentration in the base form? (d) From the ratio [base]/[acid] and the pK, calculate the intracellular pH.

23

Buffering and acid–base balance

Many factors tend to affect the acid–base status of the body. One of these is diet, and carnivores produce a more acid urine than do herbivorous mammals (except, as Claude Bernard demonstrated in 1846, when the latter are made to eat meat instead of plant material). Enzyme reactions may generate or consume hydrogen ions. Carbon dioxide production tends to acidify all the body fluids. The importance of respiration, renal excretion and of numerous pathological disturbances needs no emphasis here, but our first concern is with buffer equilibria.

It is usual to distinguish two kinds of buffer system in the body. One involves bicarbonate and dissolved carbon dioxide, with concentrations of the latter being determined by the partial pressure of carbon dioxide, which is itself controlled through pulmonary ventilation. In addition there is so-called 'non-bicarbonate buffering', which is largely due to proteins. We look first at non-bicarbonate buffering, and then at the bicarbonate/carbon dioxide system, each on its own. We then see how the two kinds of buffering work together in blood and other body fluids. Bone mineral has a long-term buffering role that is hard to quantify and is not treated here.

As explained in Chapter 21, the symbol $\{H^+\}$, rather than $[H^+]$, is used here to mean 10^{-pH}. However, there is actually very little need to use either symbol here in preference to pH.

Biomedical Calculations   Richard F. Burton
© 2008 John Wiley & Sons, Ltd

23.1 Non-bicarbonate buffering

When a small amount of HCl is added to pure water, the pH shows a marked fall. Similarly a little sodium hydroxide will cause a big rise in pH. Even the carbon dioxide from the atmosphere may markedly lower the pH of distilled water. Neutral salts like sodium chloride make no difference to the pH, but the presence of buffers tends to stabilize it. Each non-bicarbonate buffer substance buffers best over a limited pH range.

A simple buffer consists of a conjugate pair of acid and base, where an acid is a substance capable of losing a hydrogen ion to become a base and the base can take up a hydrogen ion to become an acid. In these general terms there is an equilibrium of the form:

$$\text{Acid} \leftrightharpoons \text{H}^+ + \text{Base}.$$

Here we have the examples of lactic acid, inorganic phosphate near neutral pH and creatinine:

$$C_2H_4OHCOOH \leftrightharpoons H^+ + C_2H_4OHCOO^-$$

$$H_2PO_4^- \leftrightharpoons H^+ + HPO_4^{2-}$$

$$C_4H_6ON_3H^+ \leftrightharpoons H^+ + C_4H_6ON_3$$

These examples vary with regard to the charges on the acid and base forms, but in each case the base is more negatively charged or less positively charged than the acid form. Buffering by proteins is due mainly to their imidazole and amino groups, but also, especially at low pH, to their carboxyl groups. Various other substances are used in artificial media for cells and enzymes (e.g. HEPES, Tris, boric acid/borate).

The two versions of the Henderson–Hasselbalch equation given in Chapter 22 are again relevant:

$$pH = pK + \log\left(\frac{[\text{base}]}{[\text{acid}]}\right). \tag{23.1}$$

$$\frac{[\text{base}]}{[\text{acid}]} = 10^{(pH-pK)}. \tag{23.2}$$

With these, we can do the following:

1. Given the pK value, calculate the pH of a solution from known concentrations of the acid and base forms of a buffer.

2. Given the pK value and pH, calculate the degree of dissociation of a buffer from the ratio of base to acid forms.

3. Calculate the pK from knowledge of the other variables in a particular solution.

Here are some practice examples.

23.1   A solution is made up to contain 0.1 M acetic acid and 0.2 M sodium acetate. The pK is 4.8. What is the pH? (It may help to remember log 2!)

23.2   Tris(hydroxymethyl)aminomethane, commonly called Tris, is an amine buffer. It is uncharged at high pH. Its pK is 8.3 at 20 °C. Calculate the ratio of base to acid forms at pH 7.3 and the corresponding degree of dissociation expressed as a percentage.

23.3   You may recall that the ratio $[HPO_4{}^{2-}]/[H_2PO_4{}^-]$ in arterial blood plasma of pH 7.4 is 4 (see Problem 22.9). What is the pK?

(Although naturally present in the body, phosphate is not the best of buffers in physiological extracellular media, because it tends to precipitate with calcium. However, it is an important buffer in urine.)

Buffers may be studied by titrating them *in vitro* with strong acid (usually HCl) or strong base (usually NaOH). The added amounts of acid or base are usually expressed in terms of milliequivalents rather than millimoles, mainly because this allows for the possibility that the acid could be, say, sulphuric acid, with 2 mmol of hydrogen ion per 1 mol of acid. What is shown by such a progressive titration of a simple buffer (i.e. one with only a single pK) is that the ability of a particular buffer to moderate change in pH is greatest when the pH equals the pK. In other words, the change in pH for a given addition of acid or base is least at that pH. Figure 23.1 shows how buffering ability varies with pH. It is quantified as 'buffer value' as described below.

Figure 23.1 is characteristic of simple buffers. A mixture of two buffers with very different pK values has two distinct peaks, but for a mixture of very many different buffers the peaks could in principle merge into a more or less flat-topped mountain range. Mixtures of proteins can be rather like that, with buffering ability

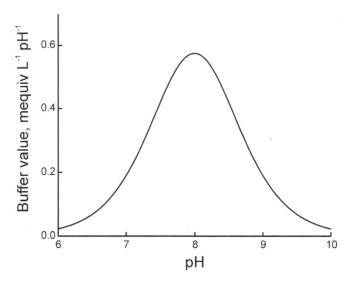

Figure 23.1 Dependence of buffer value on pH for a simple buffer with a pK of 8.0. The buffer concentration is 1 mmol/L.

varying little over a range of pH values (see Notes). Plasma proteins are generally treated as being like this.

23.4. Here is a problem that can be answered approximately just by inspection of Figure 23.1 and without much understanding of how buffering ability is quantified. A solution is buffered at pH 8.0 by a buffer with a pK of 8.0. Suppose now that the solution is made up differently, again with a pH of 8.0, but with a buffer having a pK of 7.0. By what factor would the concentration of buffer need to be increased to achieve the same buffer value?

Now we come to how buffer power is quantified. It may be expressed as the 'buffer value' or 'buffer index', defined as the quantity of strong base added per litre divided by the increment in pH that it causes. This is the same as the concentration of added strong acid divided by the resulting decrease in pH. The units of buffer value are usually given as mequiv L^{-1} pH^{-1}. These units are sometimes called 'slykes' after Van Slyke, who proposed the measure in 1922.

Here is an example. Suppose that you have 1 L of buffered solution and that you add 1 mequiv of HCl (with minimal change in volume) and that the pH falls by 0.05. The buffer value would be $(1 \text{ mequiv L}^{-1})/0.05 = 20 \text{ mequiv L}^{-1} \text{ pH}^{-1}$. Note that such a calculation is only accurate for small changes in pH, because, as Figure 23.1 shows, the buffer value varies with pH.

The curve in Figure 23.1 is calculated from the following relationship:

$$\text{Buffer value} = \frac{2.303\ C}{(2\ +\ 10^{(\text{pH}\ -\ \text{p}K)}\ +\ 10^{(\text{p}K\ -\ \text{pH})})}. \tag{23.3}$$

where C is the concentration, in mmol L^{-1}, of the buffer in its two forms (acid + base); 2.303 is ln 10.

> **23.5** When the pH equals the pK of the buffer, what is the buffer value divided by the buffer concentration C?

23.2 A link with dose–response curves

The numerical value of the peak buffer value in Figure 23.1 is 0.5756. This is suggestive of the maximum (mid-point) gradient of each of the dose–response curves in Figure 22.3 of Section 22.6, although this is given there in percentage terms as 57.56. If the gradient of any one of those curves were plotted against log(dose), the shape of the resulting bell-shaped curve would match that of Figure 23.1. Here is an opportunity to unite two seemingly disparate topics and so possibly double the usefulness of their shared ideas.

A simple and enlightening class experiment is to titrate a simple buffer from one extreme of pH to another with small increments of hydrochloric acid or sodium hydroxide and to plot the amount of acid or base added against the pH. From the changing gradient of the resulting 'titration curve' may then be derived the relationship between buffer value and pH. Provided that the buffer's pK is not too far from neutral pH, the result should resemble Figure 23.1 in shape.* If one starts from a low pH and adds sodium hydroxide, the titration curve will rise from left to right with the sigmoid shape of a dose–response curve. The mathematics is the same in the two situations. Unfortunately the analogy is spoiled if one thinks of the aliquots of sodium hydroxide as doses and the pH as the response of the glass electrode. All is well, however, if one recognizes 'log(dose)' as the logarithm of a

---

* At extremes of pH the form of the titration curve, even in the absence of buffer, reflects much more the accumulating hydrogen or hydroxyl ions.

concentration (as is often true of dose–response curves) and also equates 'amount of added base' with the ionization state of the buffer which it determines.

23.3 Bicarbonate buffering

$$pH = 6.1 + \log \frac{[HCO_3{}^-]}{0.03\, PCO_2}$$

(with $[HCO_3{}^-]$ in mmol/L and PCO_2 in mmHg)

In the box is a version of the Henderson–Hasselbalch equation to be found in many textbooks. It tells us that blood pH – meaning plasma pH regardless of the pH inside the blood cells – increases with the concentration of bicarbonate and decreases with the carbon dioxide tension.

The CO_2–HCO_3^- buffer system is fundamentally different from the others because of the way one component, the carbon dioxide, can be independently changed and regulated. It is especially important, not just to carbon dioxide transport, but because the carbon dioxide tension, and therefore the concentration of dissolved carbon dioxide, is normally adjusted in a way that maintains the arterial pH. At the same time, concentrations of bicarbonate inside and outside the cells may be altered by metabolism and are regulated by various transport mechanisms, notably in the kidneys. Here we first consider this buffer system in isolation and then see how it operates together with non-bicarbonate buffers. The reactions that concern us are these (discussed already in Section 22.1):

$$CO_2 + H_2O \leftrightarrows H_2CO_3 \leftrightarrows HCO_3{}^- + H^+$$

This is a two-stage process, with carbonic acid, H_2CO_3, in the middle. Much like the equation in the box, this tells us, with the law of mass action mind, that hydrogen ion concentrations should rise when the concentration of carbon dioxide increases and fall when the concentration of bicarbonate decreases.

The first reaction, that between carbon dioxide and water, is slow, except that it is greatly speeded up inside erythrocytes and at some other sites by the enzyme carbonic anhydrase. That is an important point to remember, but the treatment of buffering is simplified if the existence of carbonic acid is otherwise ignored. Then we can concentrate on the overall reaction at equilibrium:

$$CO_2 + H_2O \leftrightarrows HCO_3{}^- + H^+$$

The concentration of carbonic acid is in any case very low and there is no routine way of measuring it. The concentration of dissolved carbon dioxide is several hundred times that of carbonic acid. In normal arterial plasma the concentration of dissolved carbon dioxide is about 1.2 mmol/L and the bicarbonate concentration is 22–30 mmol/L.

The presence of carbonic anhydrase in our erythrocytes is not just vital to efficient carbon dioxide transport. If, like rainbow trout, we did not have it, sampled blood would not have the same pH as blood *in vivo* and the study of acid–base balance would be more complicated than it already is.

Another reaction sometimes has to be considered, although it can be ignored most of the time. This is the dissociation of bicarbonate to form carbonate:

$$HCO_3^- \rightleftharpoons H^+ + CO_3^{2-}$$

Concentrations of carbonate are generally low and unimportant to buffering, but they rise with increasing pH. One way in which the presence of carbonate is made evident is when calcium carbonate precipitates from an artificial bicarbonate-buffered solution. This may happen when much calcium is present or when the pH is allowed to rise too high. (The resulting cloudiness is accompanied by a fall in pH, as the above reaction shifts to the right.) Carbonate is ignored in the following treatment of bicarbonate buffering.

The concentrations of bicarbonate and dissolved carbon dioxide are related to pH through their own version of the Henderson–Hasselbalch equation:

$$pH = pK_1' + \log\frac{[HCO_3^-]}{[CO_2]}. \qquad (23.4)$$

The pK_1' is given its subscript to distinguish it from pK_2', which is for the second dissociation of carbonic acid, i.e. the dissociation of bicarbonate into carbonate. The concentration of dissolved carbon dioxide cannot easily be directly measured, but can be calculated as the product of the carbon dioxide tension, PCO_2, and the solubility coefficient, S, i.e. as $SPCO_2$ (see Section 17.9) A more practical version of Eq. (23.4) is therefore this one:

$$pH = pK_1' + \log\frac{[HCO_3^-]}{S.PCO_2}. \qquad (23.5)$$

At body temperature (37 °C) and physiological salt concentrations, pK_1' is 6.1 and S is 0.03 mmol/(L mmHg). Under these conditions and using the same units,

Eq. (23.5) may therefore be written as follows, the formula in the box above:

$$pH = 6.1 + \log \frac{[HCO_3^-]}{0.03 \, PCO_2}. \tag{23.6}$$

23.6   Especially if you prefer to work with kPa rather than mmHg, amend Eq. (23.6) accordingly.

If the PCO_2 is 40 mmHg, as in arterial plasma, $0.03 \, PCO_2$, equal to $[CO_2]$, is 1.2 mmol/L. What should $[HCO_3^-]$ be for the pH to be 7.4? Here $(pH - pK_1')$ is $(7.4 - 6.1) = 1.3$. As stressed in Chapter 18, 0.3 is log 2, so that 1.3 is log 20. Equation (23.4) can therefore be written as:

$$(pH - pK_1') = \log 20 = \log \frac{[HCO_3^-]}{1.2 \, mmol/L}.$$

Evidently $[HCO_3^-]/1.2$ mmol/L must equal 20 in this situation, so that $[HCO_3^-]$ is 24 mmol/L. That is within the range of normal values given above (22–30 mmol/L). So now we have a consistent set of normal values to keep in mind for pH, $[HCO_3^-]$ and PCO_2, namely 7.4, 24 mmol/L and 40 mmHg, respectively. Of course there is variation around these values.

It is possible to simplify Eq. (23.6). Thus

$$pH = 6.1 - \log 0.03 + \log \frac{[HCO_3^-]}{PCO_2}.$$

$$= 7.62 + \log \frac{[HCO_3^-]}{PCO_2} \tag{23.7}$$

This is slightly more convenient, but textbooks do not give this, perhaps because it obscures the reasoning behind the usual equation (and complicates unit analysis!). (Combining pK_1' and S into one constant can be helpful when one is studying body fluids, especially those of other animals, for which the separate values of pK_1' and S are not already known.)

Some books give a non-logarithmic version of Eq. (23.7):

$$[H^+] = K \cdot \frac{PCO_2}{[HCO_3^-]} \tag{23.8}$$

With $[H^+]$ in nmol/L, K at 37 °C is 24 nmol mmol L^{-2} mmHg^{-1} or 180 nmol mmol L^{-2} kPa^{-1}. Perhaps because the units are cumbersome, they are often unstated. There is an obvious appeal to Eq. (23.8), although one should bear in mind the distinction between [H$^+$] and {H$^+$}. The normal arterial value for {H$^+$} is $10^{-7.4} = 4 \times 10^{-8}$ mol/L = 40 nmol/L.

All that has been said so far applies irrespective of the presence of non-bicarbonate buffers. We come to these shortly, but an important point must be established first. In a solution gassed with a CO$_2$–O$_2$ mixture and containing bicarbonate and other inorganic ions, but no non-bicarbonate buffers, the concentration of bicarbonate is virtually independent of the carbon dioxide tension. Some people have difficulty accepting this, because the pH obviously is affected. Consider what happens when the carbon dioxide tension is raised. Hydrogen ions are generated in the following reaction and the pH falls:

$$CO_2 + H_2O \rightarrow HCO_3^- + H^+$$

If the reaction is moving to the right, the bicarbonate concentrate must indeed rise, but the point is that it does so imperceptibly. Suppose that the pH falls from 7 to 6. If {H$^+$} is approximated as [H$^+$], this means that the latter rises from 10^{-4} mmol/L (10^{-7} mol/L) to 10^{-3} mmol/L (10^{-6} mol/L). This is a rise of 0.0009 mmol/L. The bicarbonate concentration rises by the same amount, say from 25.0000 mmol/L to 25.0009 mmol/L. This change is far too small to be measured and this is why I wrote 'the concentration of bicarbonate is *virtually* independent of the carbon dioxide tension'. As discussed below, the situation is entirely different when non-bicarbonate buffers are present. (I have simplified the argument by ignoring changes in the concentration of hydroxyl ions, but these are present in minute concentrations too.)

23.4 CO$_2$/HCO$_3^-$ and non-bicarbonate buffers together

Body fluids contain both bicarbonate and non-bicarbonate buffers and so too do many artificial media for cells or tissues. A point made earlier is that, in the absence of non-bicarbonate buffers, a change in PCO$_2$ produces virtually no change in bicarbonate concentration. This is emphatically untrue when other buffers are present. Suppose that the PCO$_2$ rises in a solution containing bicarbonate and some other buffer which, for the sake of illustration, we may take as inorganic phosphate. The following reactions occur:

$$CO_2 + H_2O \rightarrow HCO_3^- + H^+$$
$$H^+ + HPO_4^{2-} \rightarrow H_2PO_4^-$$

or, combining these,

$$CO_2 + H_2O + HPO_4^{2-} \rightarrow HCO_3^- + H_2PO_4^-.$$

Now a rise in PCO_2 does produce a significant rise in bicarbonate concentration as $[HPO_4{}^{2-}]$ falls. In a solution studied *in vitro* there would be no change in the concentrations of other ions. Then, in accordance with the principle of electroneutrality (Section 9.1), the total concentration of these buffer anions, in mequiv/L, would have to stay constant. With square brackets denoting concentrations in mmol/L, their total concentration in mequiv/L is:

$$[HCO_3^-] + 2[HPO_4^{2-}] + [H_2PO_4^-].$$

To put this more generally, in a system closed for everything except carbon dioxide, the total net charge on all the buffers present must stay constant. This is an important principle in the quantitative treatment of buffering.

As just noted, the total net charge on buffers stays constant if there are no changes in the other ions present. However, the balance of those other cations and anions may alter – with significant consequences for acid–base balance. Table 23.1 shows the concentrations of cations and anions in fairly typical, but simplified, extracellular fluid. They are expressed in mequiv/L. The total concentrations of each are identical. In accordance with the principal of electroneutrality, the bicarbonate concentration must equal the sum of all the cations minus the sum of all other anions, namely 27 mequiv/L. One sees here, not just the general interdependence of ion concentrations, but specifically the dependence of bicarbonate concentration on all other ions. For example, if chloride is depleted as a result of persistent vomiting of gastric acid, an increase in plasma bicarbonate makes up for part of the potential discrepancy in the balance sheet. The resulting tendency for pH to rise (assuming that PCO_2 does not rise proportionately) leads to increased

Table 23.1 Concentrations of cations and anions in a solution similar to plasma, expressed in mequiv/L

Cations	(mequiv/L)	Anions	(mequiv/L)
Na$^+$	144	Cl$^-$	104
K$^+$	4	HCO$_3^-$	27
Free Ca^{2+}	3	Protein^{z-}	20
Free Mg^{2+}	1	Lactate$^-$	1
Total	152	Total	152

dissociation of hydrogen ions from proteins, thus raising their net negative charge and their anionic contribution.

We see therefore that bicarbonate cannot be adjusted independently of other ions as a means of regulating pH. Mechanisms for adjusting bicarbonate always involve other ions.

While a full ionic balance sheet is invaluable in interpreting acid–base disturbances, there is generally less information available than there is in Table 23.1. This is why the anion gap and strong ion difference (SID) of Section 9.3 are seen as informative. In this example the anion gap, calculated as $\{([Na^+] + [K^+]) - ([Cl^-] - [HCO_3^-]\}$ is 17 mequiv/L and, as usual, is due mainly to protein. However, as already noted in Section 9, it can be raised in a variety of pathological conditions with an accompanying fall in bicarbonate concentration (metabolic acidosis). The SID for the data of Table 23.1, calculated from its simplified version as $([Na^+] + [K^+] - [Cl^-])$, is 44 mequiv/L. In normal plasma it is mainly due to bicarbonate and protein. Although a high SID may therefore reflect a high bicarbonate concentration, it could also be due to abnormally high concentrations of organic anions, perhaps coupled with diminished bicarbonate levels. It is not a good indicator of acid–base status on its own, therefore.

23.5 The whole body: diet and the titratable acidity of urine

The principal of electroneutrality must apply to the whole body too, though that is something that is neither practically nor usefully tested. What can be measured is the net or base input or output in terms of diet and urine composition (see Notes). Assessing the dietary contribution to acid–base balance is hardly a routine procedure as it involves ashing food samples and either titrating the ash or analysing its elemental content. The urine, however, is easily analysed and its influence on whole-body acid–base balance is measurable, in part, as its 'titratable acidity'. This may be positive or negative. Various definitions and measurement procedures have been used for this, giving different answers, but all involve titrating samples with NaOH or HCl, depending on whether they are initially acid or alkaline. The body excretes much of its excess acid as ammonium ions (NH_4^+) and the amount of this in the urine should be added to the titratable acidity. In principle, the content of organic anion should be subtracted, though this is generally less important.

More easily determined is the urine anion gap, $([Na^+] + [K^+] - [Cl^-])$, that is mentioned in Section 9.3. To understand this, we may think of the urine in simplified terms and equate cations and anions as follows, with 'Cr-H$^+$' standing for hydrogen ions buffered by creatinine.

$$[\text{Na}^+] + [\text{K}^+] + 2[\text{Ca}^{2+}] + 2[\text{Mg}^{2+}] + [\text{NH}_4^+] + [\text{Cr-H}^+]$$
$$= [\text{Cl}^-] + [\text{HCO}_3^-] + 2[\text{SO}_4^{2-}] + [\text{H}_2\text{PO}_4^-] + 2[\text{HPO}_4{}^{2-}].$$

By rearrangement:

$$[\text{Na}^+] + [\text{K}^+] - [\text{Cl}^-] = [\text{HCO}_3^-] + 2[\text{SO}_4^{2-}] + [\text{H}_2\text{PO}_4{}^-]$$
$$+ 2[\text{HPO}_4{}^{2-}] - 2[\text{Ca}^{2+}] - 2[\text{Mg}^{2+}] - [\text{NH}_4{}^+] - [\text{Cr-H}^+].$$

From this it is apparent that the anion gap is decreased in acid urine, because less bicarbonate is excreted and more hydrogen ion is buffered by ammonia, phosphate and creatinine.

23.6 Other aspects of acid–base balance

There are important aspects of this very quantitative topic that can be dismissed briefly on the grounds that there are few requirements or useful possibilities for simple calculation. These include acid–base graphs of various kinds that are invaluable aids to learning, diagnosis and the assessment of treatment (see Notes). The Siggaard–Andersen curve nomogram, using measurements of $P\text{CO}_2$, arterial blood pH and haemoglobin, provides precise information on plasma bicarbonate and other relevant variables, namely buffer base, standard bicarbonate and base excess, but involves no calculation. Base excess is the amount of acid or base which would restore 1 L of blood to a normal acid–base state at a $P\text{CO}_2$ of 40 mmHg. It is positive in metabolic alkalosis and negative in metabolic acidosis. The base excess provides a guide to how much acid or base, per kilogram of body mass, it may be appropriate to infuse in the treatment of those two disorders. However, the whole body is more complicated than blood, so that reckoning the dose involves judgement, timing and inexact rules of thumb that contrast with the precision of the nomogram itself.

   The latter parts of this chapter are barely represented in the problems that follow. Calculations on the anion gap and SID would require only simple addition and subtraction. Some of the other quantitative aspects of acid–base balance are not usefully approached through simple calculations.

Normal values for arterial plasma consistent with

$$pK'_1 = 6.1$$

and

$$S_{CO2} = 0.030 \, L^{-1} \, mmHg^{-1} = 0.226 \, L^1 \, kPa^{-1}$$

include:

$$pH = 7.4$$

$$[HCO_3{}^-] = 24 \, mmol/L$$

$$[CO_2] = 1.2 \, mmol/L$$

$$PCO_2 = 40 \, mmHg = 5.3 \, kPa$$

Problems

23.1   See Section 23.1.

23.2   See Section 23.1.

23.3   See Section 23.1.

23.4   See Section 23.1.

23.5   See Section 23.1.

23.6   See Section 23.3.

23.7   Do buffers always stabilize pH? A solution of HEPES buffer is made up with a pH of 7.4 at 20 °C. (Its pK at 20 °C is 7.55.) Warmed to 37 °C, its pH is found to be 7.16. Explain.

23.8   A solution is buffered at pH 7.4 with 20 millimolar HEPES. The temperature is 20 °C so that the pK is 7.55. (a) Calling the two forms B and BH$^+$, calculate the ratio $[B]/[BH^+]$ and then both $[B]$ and $[BH^+]$. (b) HCl at 4 mM is added to the solution. What is the resulting pH? (c) From (b), what is the approximate buffer value at pH 7.2 (i.e. midway between pH 7.4 and 7.0)? (d) What is the buffer value at pH 7.2 calculated more accurately from Eq. (23.3)?

23.9   Taking the solubility coefficient of carbon dioxide as 0.03 mmol L^{-1} mmHg^{-1} and pK'_1 as 6.1, calculate the pH of a solution containing 30 mM bicarbonate that is equilibrated at $37\,^{\circ}C$ with carbon dioxide at a partial pressure of 35 mmHg.

23.10 A man is admitted to hospital in a semicomatose condition, a diabetic who has not been taking his insulin. His arterial pH is 7.10, coupled with a PCO_2 of 20 mmHg. Calculate the corresponding bicarbonate concentration.

23.11 A man with pneumonia has an arterial pH of 7.19 and PCO_2 of 11.5 kPa. Calculate the corresponding bicarbonate concentration.

23.12 A solution contains bicarbonate, but no other buffers. Its initial pH is 7.39. (a) What is the pH following doubling of the PCO_2? (b) What is the pH if the PCO_2 stays the same and the bicarbonate concentration is doubled?

23.13 If the normal range of bicarbonate concentration in arterial blood were, say, 22–30 mM, what would be the corresponding difference in pH values in the hypothetical situation that the PCO_2 stays constant? What does the answer imply, given that the pH normally varies between about 7.35 and 7.45?

23.14 Imagine that each of the following substances is metabolized as far as possible to carbon dioxide and water leaving, in two cases, some inorganic ions (K^+ or SO_4^{2-}). In each case, what would be the final effect on acid–base balance? (Ignore the carbon dioxide that is produced.) (a) Potassium citrate; (b) citric acid; (c) cysteine or methionine (sulphur-containing amino acids).

Appendix A
Basic mathematics and mathematical language

Numbers

'*Integers*' are whole numbers. Thus 345 is an integer and 345.1 is not.

'*Scientific notation*' combines small numbers (with one digit before the decimal point) and powers of 10, for example 3.451×10^2 in place of 345.1. Note the convenience of expressing 0.0000078 in scientific notation as 7.8×10^{-6}. Especially with handwritten numbers, avoid starting with a decimal point, i.e. write '0.17', not '.17'.

The word '*recurring*' is applied to digits after the decimal point that repeat forever. Thus $2/3 = 0.66666666\ldots$ This may be written as $0.\dot{6}$, with a dot over a single six. *Euler's number*, or the exponential constant, e, has the value $2.71828\ldots$

'*Dimensionless numbers*' are pure numbers, without units.

Precision

Dictionaries may say that this is the same as accuracy, but a figure may be given with great precision, but little accuracy: '5' is less precise than '5.0', because '5' may just mean '5 as the nearest whole number', or 'between 4.5 and 5.5'. Thus, one may speak of the cardiac output as being (commonly or approximately) 5 L/min, but at a particular time in a particular individual it could be more precisely measured as 5.0 L/min. So, in general, do not omit the zero if greater precision is intended.

Biomedical Calculations   Richard F. Burton
© 2008 John Wiley & Sons, Ltd

That said, it is not uncommon for '1', say, to mean 'exactly 1'; this is usually clear from the context, especially when only integers are possible.

Calculators give answers to a very high degree of precision (to many '*places of decimal*', referring to the numbers of digits after the decimal point). However, the results of a calculation should be recorded only to a sensible degree of precision. Judge that from the accuracy or precision of the numbers put into the calculation; an answer is unlikely to be more accurate than the numbers it is based on.

Reducing the number of decimal places involves either *rounding up* or *rounding down* – so that the revised number is as close as possible to the original. Thus, to one place of decimal, 1.36 becomes 1.4 and 1.34 becomes 1.3. 1.35 is half-way between 1.3 and 1.4, but the convention is to round fives upwards, i.e. to 1.4 in this case.

When numbers are carried forward in a series of calculations, some excess of significant figures should be maintained until the end, however, so that small roundings do not accumulate to become a significant error.

Precision may be specified in terms of numbers of '*significant figures*': 5023, 49.96 and 0.6754 are all given to four significant figures; 5000, 50 and 0.7 are those same three quantities expressed to one significant figure.

Quantities may be referred to very imprecisely (but often usefully) in terms of '*orders of magnitude*' that are usually thought of as relating to factors of 10. You might guess that the mass of a mouse is roughly 15 g, meaning nearer to 15 g than to either 1 g or 100 g. You could then say that a mouse is about four orders of magnitude lighter than yourself. (If you are 50–200 kg, then this is something like 10 000 times the roughly 15 g of the mouse.) There are no defined cut-off points for orders of magnitude.

Operations

An 'operation' is the action of adding, subtracting, multiplying, dividing, taking square roots, etc. Corresponding 'operators' are $+$, $-$, $\times$, $\div$ and $\sqrt{}$.

Multiplication may be represented in several ways:

most usually	2×4
in printed text	$2 \cdot 4$, but beware: the dot can be confused with a decimal point
on computer keyboard	$2 * 4$.

In addition, two or more numbers, symbols or expressions may be written side by side,

$$\text{e.g. } 2A, \ AB \text{ or } 3(A + B)C.$$

Division may also be represented in various ways:

$$2 \div 5, \ 2/5, \ \frac{2}{5}.$$

Expressions

These are combinations of numbers (or algebraic symbols) and operators, of which we have already met a few simple ones, e.g. 2×4 and $2 \div 5$. There may be alternative ways of writing expressions, as illustrated next.

Square

Usually $4^2 = 16$
computer keyboard $4\textasciicircum 2 = 16.$

Square root

$$\sqrt{4} = 2$$

also written as $4^{1/2}$ and $4^{0.5}$

and, on computer keyboard, as $4\textasciicircum 0.5$.

Reciprocal

The reciprocal of a number is that same number divided into 1 ('1 over that number'). Thus 1/7 is the reciprocal of 7. It is also written as 7^{-1}.

Numerator, denominator, quotient

A fraction or ratio is written with a numerator and a denominator, the first on the top line and the second on the bottom. Thus:

$$\frac{\text{Numerator}}{\text{Denominator}} = \frac{\text{top line}}{\text{bottom line}}$$

A quotient is the result of dividing one number by another. Thus the quotient of 15 divided by 3 is 5.

Indices, exponents, powers

In the expression x^a, the a is called an 'index' or 'exponent'. The expression is pronounced 'x to the power a', or 'x to the a'. When a is a positive integer (whole number), x^a is simply the product of a x's, so that, for example,

$$x^3 = x \times x \times x \text{ (i.e. 'x cubed').}$$

(To say that x^a is the product of a x's no longer makes sense, however, when a is a fraction, negative or zero.)

We have already noted that $4^{0.5} = \sqrt{4}$ and that $7^{-1} = 1/7$. Likewise $7^{-2} = 1/7^2$ and $7.8 \times 10^{-6} = 7.8/1\,000\,000$. Note that $x^0 = 1$, whatever the value of x.

You may check that negative, fractional and zero indices do make sense in relation to rule 1 that follows.

Rules relating to indices

Indices can be manipulated using the following rules.

1. $x^a \times x^b = x^{a+b}$ (also written as $x^{(a+b)}$).

2. $x^a / x^b = x^{a-b}$.

3. $1/x^b = x^{-b}$.

4. $(x^a)^b = x^{ab}$.

5. $x^a \times y^a = (xy)^a$.

6. $x^0 = 1$.

These are not independent rules, but follow one from another. Thus rule 3 can be obtained from rule 2 by taking $a = 0$. Rule 6 follows from rule 2 with $a = b$.

If you are unsure whether you remember these rules correctly, test them with actual numbers. For example, as a test of rule 1:

$$3^2 \times 3^3 = (3 \times 3) \times (3 \times 3 \times 3) = 3^5 = 3^{2+3}.$$

e^a is sometimes written for convenience as $\exp(a)$.

Order of operations

Consider the expression: $10 \div 5 + 2^3$. In what order do you carry out the operations? The correct procedure is to divide 10 by 5 and then add the cube of 2 (answer $= 10$). You get the wrong answer if, for example, you divide 10 by the sum of 5 and 2, then cube the result. The rule in such expressions is: *Powers first* (cubing in this case), then *multiplication and division*, then *addition and subtraction*.

If the operations need to be performed in an order different from that, then this is made clear by the use of *brackets*. The rule then is that expressions inside brackets are evaluated first. Thus,

$$10/(5 + 2^3) = 10/13 = 0.77.$$

As another example,

$$(9 + 1)^2/(3 + 2) = 10^2/5 = 20.$$

It is possible to have pairs of brackets nesting one (or more) inside another. Thus:

$$((9 - 1) + (2 + 3)) \times 2 = (8 + 5) \times 2 = 13 \times 2 = 26.$$

When there are brackets embedded within brackets, work outwards from the inside, evaluating expressions in the innermost pairs of brackets first.

There are alternative kinds of bracket, notably (), [] and {}. They generally mean the same thing. The previous example is made clearer by using two kinds:

$$[(9 - 1) + (2 + 3)] \times 2.$$

Multiplication may be indicated by having brackets and numbers placed side by side:

$$2(2 + 3)(5 - 1) = 2 \times 5 \times 4.$$

Aside from appearance and economy, there is no harm in using superfluous brackets – and sometimes they can be helpful. Thus, it is permissible, if pointless and clumsy, to write:

$$10/5 + 2^3 \text{ as } \{(10/5) + ([2]^3)\}.$$

None of these brackets affect this expression, but the outer pair, i.e. {}, may help to highlight the whole expression as an entity, especially when it is embedded in text. Make sure that you close as many brackets as you open!

Remember that square brackets are also used for concentrations, so that [Na] means 'concentration of sodium'.

Rearranging expressions

Expressions may be rearranged so long as attention is paid to the above rules. Obviously $(3 + 2)$ is the same as $(2 + 3)$, and 3×7 is the same as 7×3. Similarly, $5 - 2$ is the same as $-2 + 5$.

As examples other kinds of rearrangement, note these:

$$2(5 + 4) = 2 \times 5 + 2 \times 4.$$

$$7/9 + 5/9 = (7 + 5)/9.$$

$$2(5 + 4) + 3(5 + 4) = (2 + 3)(5 + 4).$$

$$\frac{5}{6} + \frac{7}{9} = \frac{5 \times 9 + 7 \times 6}{6 \times 9}.$$

These examples may seem less trivial when the numbers are replaced with algebraic symbols, e.g. $a(x + y) = ax + ay$.

Equations and formulae

Equations are statements that one expression is equal to another expression or quantity, for example '$2 \times 4 = 64^{0.5}$' or, as just above, '$2(5 + 4) + 3(5 + 4) = (2 + 3)(5 + 4)$'. Usually, however, the word is applied to equations involving symbols, such as '$A^2 = 64$', for which the solution is $A = 8$. A 'solution' is a value (in this case the value of A) for which the equation is true.

The word 'formulae' is applied to equations involving symbols that allow you to calculate specific values of a particular quantity by replacing symbols with known values, e.g. the formula for calculating the volume of a sphere, V, from its radius, r, which is '$V = (4/3)\pi r^3$'.

Rearranging equations

The simplest rule to remember is this, that if you have two expressions that are equal (including the two sides of an equation) and you do the same thing to both, then they are still equal. Here are some examples.

Subtracting 3:	$y = x + 3$	becomes	$y - 3 = x.$
Adding x:	$y - x = 7$	becomes	$y = x + 7.$
Multiplying by 5:	$y = x/5 + k + 2$	becomes	$5y = x + 5k + 10.$
Dividing by a:	$ay = x + 2a$	becomes	$y = x/a + 2.$
Squaring:	$\sqrt{y} = x + 1$	becomes	$y = (x + 1)^2.$
Taking square roots:	$(x + y)^2 = 9$	becomes	$x + y = 3.$
Inverting:	$1/y = a/(b + 3)$	becomes	$y = (b + 3)/a.$

If you are not used to this, it may help to write out every step, especially with more complicated equations. In the case of the first example, subtracting 3, you would then write:

$$y = x + 3$$

$$y - 3 = x + 3 - 3$$

$$y - 3 = x.$$

Notice that in the first two examples the 3 and the x were effectively moved from one side to the other *with a change of sign*. This can be taken as a general procedure, but with more complicated expressions you need to watch out that what you are doing is valid.

When you multiply or divide both sides of an equation by the same quantity, as in the third and fourth examples just above, the procedure is called '*multiplying through*' or '*dividing through*'.

When an equation has a top and a bottom line on each side, you can multiply through so that neither side has a bottom line. This is '*cross-multiplying*'. Thus, with both sides multiplied by $(a + 3)$ and by 5:

$$\frac{y}{a + 3} = \frac{x + 7}{5} \qquad \text{becomes} \qquad 5y = (a + 3)(x + 7).$$

More complicated rearrangements should be checked. For guidance on this, see Sections 11.6 and 11.13.

More symbols

$\pm$	'plus or minus'
$\approx, \doteq$	'approximately equal to'
$\neq$	'(is) not equal to'
$>$	'(is) greater than'
$<$	'(is) less than'
$\geq$	'(is) greater than or equal to'
$\leq$	'(is) less than or equal to'
$\gg$	'(is) much greater than'
$\ll$	'(is) much less than'
$\propto$	'proportional to'
∞	'infinity'
Δ	'difference between'. Thus Δx means the difference between two values of x. If these are. x_1 and x_2, $\Delta x = (x_1 - x_2)$. (See 'Subscripts', below.)
Σ	'sum of'. Thus Σx means 'the sum of all values of x'.
$\therefore$	'therefore'

Symbols for quantities

These are usually letters of the alphabet, including Greek letters. It helps to use initial letters, e.g. T for temperature, t for time, W for work. X and Y, or x and y, often stand for unknown quantities. Respiratory physiology has an agreed set of symbols. Italics are used in these examples, as is common in printed text.

Symbols, like any abbreviations, *should always be defined*. Avoid symbols that have special mathematical meanings, e.g. Δ, Σ.

Subscripts

As in the examples x_1 and x_2 above, the subscripts signify different values of x. Thus t_0 might stand for 'time zero'. Again it may help to use initial letters. Thus, for example, the subscripts A and V may indicate arterial and venous values of some property of blood. Subscripts may be more complicated, as in P_{CO_2}, but the idea is the same. Such subscripts may be regarded as adjectives attached to nouns. A combination of symbols, like $(P_{CO_2})_V$, meaning 'tension of carbon dioxide in venous blood', may look complicated, but it can be manipulated mathematically as a single entity, just like a simple 'x'.

Appendix B
Some non-metric units

This table gives the metric (SI) equivalents of some non-metric measures of distance, volume and mass. The right-hand column shows the reciprocals of the respective conversion factors (so that, to take the first example, 1 mm = 0.0394 in).

In some cases, as shown, there are differences between the British ('imperial') and US systems. The US system has separate liquid and dry capacity measures, of which only the liquid measures are shown.

Non-metric	Metric equivalent	Reciprocal
Distance		
1 inch (in)	25.40 mm	0.0394
1 foot (ft)	0.305 m	3.281
1 yard (yd)	0.9144 m	1.0936
1 mile (mi)	1.609 km	0.6214
Volume (liquid)		
1 minim, imperial (min)	0.05919 mL	16.89
1 minim, US (min)	0.06161 mL	16.23
1 pint, imperial (pt)	0.5683 L	1.760
1 liquid pint, US (liq pt)	0.4732 L	2.113
1 gallon, imperial (gal)	4.546 L	0.2200
1 gallon, US (gal)	3.785 L	0.2642

(*Continued*)

Biomedical Calculations   Richard F. Burton
© 2008 John Wiley & Sons, Ltd

Non-metric	Metric equivalent	Reciprocal
Mass		
1 grain (gr)	0.0648 g	15.43
1 dram	1.772 g	0.564
1 ounce (oz)	28.35 g	0.0353
1 pound (lb)	0.4536 kg	2.2046
1 imperial ton/US long ton	1.016 tonne	0.984
1 US short ton	0.9072 tonne	1.102

Appendix C
Notes

1.4 When to specify substances along with units

For good practice in theses and publications, see Ebel *et al.* (2004).

1.8 Dimensional analysis

For an interesting account of dimensional analysis in biology, see McMahon and Bonner (1983).

4.2 Specific volume

The partial specific volume of the plasma proteins (albumin plus globulin) is as given by Van Slyke *et al.* (1950), who calculated protein concentrations as 6.25 times the nitrogen concentration (measured by the Kjeldahl method). The density of protein-free plasma, 1.007 g/mL, is also their value, except that they expressed it as specific gravity. Their optimum formula for calculating P from D is:

$$P = 373(D - 1.007).$$

They calculated the partial specific volume of the proteins from this relationship, so that the calculation given here is the reverse of theirs.

   The formula is invalidated if there is much lipid in the plasma.

Biomedical Calculations   Richard F. Burton
© 2008 John Wiley & Sons, Ltd

4.4 Thinking about a formula

Equation (4.1) is that of Siri (1961).

6.2 'Let x be the unkown'

Here, in two versions, is another way of solving the second problem (concerning the 5 mM solution of KCl).

$$5 \text{ mmol/L} = 5 \text{ mmol/L} \times 1 \text{ mol/1000 mmol} = 0.005 \text{ mol/L.}$$

$$0.005 \text{ mol/L} \times 74.6 \text{ g/mol} = 0.373 \text{ g/L.}$$

$$0.373 \text{ g/L} \times 250 \text{ mL} \times 1 \text{ L/1000 mL} = 0.09325 \text{ g.}$$

For clarity, the calculation was broken up into stages, but these can be combined as follows:

$$5 \text{ mmol/L} = 5 \text{ mmol/L} \times 1 \text{ mol/1000 mmol} \times 74.6 \text{ g/mol} \times 250 \text{ mL}$$

$$\times 1 \text{ L/1000 mL}$$

$$= 0.09325g \text{ in 250 mL.}$$

8 Calculating drug doses

The so-called 'dimensional analysis' of dose calculation is usefully reviewed by Arnold (1998), who also discusses unsatisfactory aspects of other current methods.

9.4 Osmoles and osmolality

Human physiology is not much concerned with actual osmotic pressures expressed in pressure units. This is despite the importance of the osmotic gradients, often very small, that govern the diffusive movements of water within the body. Nevertheless, it is useful to have some idea of their magnitude. At 0 °C and a pressure of 1 atmosphere, 1 mol of gas occupies 22.4 L (Chapter 17). Surprisingly, perhaps, it is equally true that 1 mole of solute in 22.4 L water gives an osmotic pressure of 1 atm. Those statements are only exactly true of ideal gases and solutions and it is better to say that 1 *osmole* of solute in 22.4 L water gives an osmotic pressure of 1 atm. This allows us to calculate the osmotic pressure of typical body fluids. Taking a concentration of 290 mosmol/L water we may calculate as follows:

$$1 \text{ atm} \times 290 \text{ mosmol/L} \times 1 \text{ osmol}/1000 \text{ mosmol} \times 22.4 \text{ L}$$

$$= 6.50 \text{ atm. This is } 4940 \text{ mmHg.}$$

This value applies at $0\,°C$ (273 K). At $37\,°C$ (310 K) the osmotic pressure is higher in proportion to the absolute temperature, i.e. $4940 \times 310/273 = 5610$ mmHg. However, the important point to note is less its actual value than that it is very high compared with any hydrostatic pressures within the body.

Textbooks of physical chemistry give fuller accounts. These involve the gas constant R (Section 17.4), but why that should be relevant to solutions is something you may not discover in a shelf-full of such books. Perhaps their authors agree with van't Hoff, who wrote in 1885 that it is obvious. Walter Nernst gave a proof in his textbook of 1893 for the situation where the osmotic pressure is due to dissolved gas. Sometimes ancient books tell us things that newer ones have no space for.

Note that concentrations are expressed in terms of volumes of water and not of solution. Here 1 L of water can be taken as 1 kg and, as is confirmed experimentally, it is osmolal concentrations that matter, not osmolar concentrations.

The osmolality of a solution relates exactly to the colligative properties of freezing point depression and vapour pressure lowering. However they are calibrated, osmometers measure one or other of these. A solution containing solutes at a total concentration of 1 osmol/kg water has a freezing-point depression of $1.858\,°C$. Concentrations of body fluids have often been expressed simply as $°C$ of freezing point depression, especially by comparative physiologists.

9.5 Osmolar gap

Dorwart and Chalmers (1975) have compared different formulae, including Eq. (9.2), for estimating serum osmolality from the concentrations of individual serum constituents.

9.7 Intracellular osmosity gap and anion gap

For further discussion of these in the context of animal cells generally, and also information on specific intracellular compounds responsible for osmosity gaps, see Burton (1983b, c).

9.10 Free and bound concentrations, activities

Activity may be thought of as that property of a solute which, when substituted for the concentration in some equilibrium function, makes the latter thermodynamically exact.

For more on ion pairs, see Burton (1983a) and Levine (2003).

10.11 On the different kinds of formulae and equations

Here is how the data of Figure 10.4 were obtained. First a set of random numbers, 'C', was generated in a computer spreadsheet. Values of A and B were then obtained respectively as $(3R_1C + 1)$ and $2R_2C$, where R_1 and R_2 are other random numbers, each set having a mean of 1.

A useful paper on the difficulties of finding true functional relationships by statistical methods is that of Warton *et al.* (2006).

11.4 Distortion of erythrocytes in passing along narrow blood vessels

The calculation of the limiting diameter is essentially that of Canham and Burton (1968). Assuming different dimensions for the erythrocyte than those used here, they obtained an average diameter of 3.3 μm. Intermediate values have been found in other studies. In fact there is no one definitive value, because erythrocyte dimensions change with their age and vary within a given blood sample. More important is the question of whether erythrocytes can actually distort as postulated, so it is reassuring that the minimum diameter of pore through which erythrocytes can pass, in a filter, is near 2.6 μm (Reinhardt and Chien, 1985).

11.14 Generalizing the formula to the human body

The justification for applying Eq. (11.16) to the human body is discussed more fully by Burton (2008). For further discussion, see the Notes for 11.16.

11.15 Surface/volume and surface/mass ratios

Despite the huge literature on Kleiber's rule, extending over more than a century, it has minor practical significance to human physiologists and clinicians, except inasmuch as they rely on experiments using smaller species. This is because the variation in body mass amongst adult humans is small compared with other aspects of variability and because development in children involves more than just an increase in mass. A wide range of body mass is needed to establish the proper exponent and that is provided by the variety of mammalian species, from shrew to whale. However, mammals vary in more ways than size and the potential effects of these on metabolic rates must be considered too. White and Seymour (2003) took account of variations in body temperature, digestive state and phylogeny and then found no support for the $^3/_4$ exponent. Indeed their data on 619 species varying in mass over five orders of magnitude show that basal metabolic rate varies with body mass$^{2/3}$.

The disgraceful story of the elephant is recounted by West *et al.* (1962).

11.16 The surface area of the human body

The mathematical relationship between Eqs (11.16) and (11.8) is explained by Burton (2008). (The algebra involves a Taylor series expansion.) The parameters of these equation are linked, but awkwardly. The constants α' and β' in Eq. (11.16) are in principle definable for individuals, though typical values have yet to be established. They may also be defined statistically for population samples as regression coefficients, as is true of K, a and b in Eq. (11.18). Values are sometimes then incompatible with those that could possibly apply to individuals, this being the distorting effect of regression analysis discussed in Section 10.11. This does not invalidate either type of equation when used for the prediction of surface areas and there is therefore no reason to abandon Eq. (11.18) in the usual practical contexts. However, Eq. (11.16), having a theoretical basis, should be superior in studies of factors such as obesity that may influence the dependence of surface area on mass and height.

11.17 Standard formulae for body surface area

Burton (2008) summarizes values of K, a and b in Eq. (11.18) that have been obtained in different studies.

12.1 Variations in fat-free mass with height and age

There are many other published data that accord with the relationship shown in Figure 12.1.

12.2 The Rohrer index, or 'height–weight index of build'

Rohrer described this index in 1908. The 'height–weight index of build' is the name used by Bardeen (1920).

12.3 The body mass index; estimating body fat from body mass and height

The arguments and method of modelling are more fully presented by Burton (2007b).

The difficulties of establishing true statistical relationships are discussed by Warton *et al.* (2006).

The exponent for height in Eq. (12.2) (i.e. 2.05) may be found either by regressing log(mass) on log(height) or by nonlinear regression. Quetelet's account of 1835 is also available in his book of 1869.

Despite its statistical basis, some people regard the BMI as more a predictor of absolute fat content (i.e. as mass) than of percentage fat. They do so because the BMI does actually correlate more strongly with fat mass than with percentage fat. However, this is at least to some extent inevitable, given that, for a constant percentage fat content, both fat mass and the BMI increase with body mass. What matters practically is that, if you estimate fat mass from BMI and then convert the result to a percentage, you do not obtain a better estimate of that.

12.5 Weightlifting – and the cross-sectional area of muscle

Lietzke (1956) was the first to use weightlifting data in this way, and his data points give an excellent fit to theory. Amongst useful later papers on the topic is that of Ford *et al.* (2000). The latter authors put more emphasis on height, finding that weight lifted varied almost exactly with height squared (actually height$^{2.16}$), implying that muscle mass varies with height cubed.

12.6 Estimating body fat from skinfold thickness measurements

Evidence for the validity of Eq. (12.8) has been obtained by Burton and Cameron (2008).

Problem 13.5

The origin and definition of the 'standard 70 kg man' are hard to pin down (Burton, 2000), although he is well known to physiologists. Keys and Brožek (1953) refer to 'our standard "normal young man"', giving details of his composition. A 'standard man' of 70 kg is described in Supplement 6 to the *British Journal of Radiology* (1955). He is intended to provide a common basis for calculations in radiological laboratories. I presume him to have been typical of medical-student subjects at one time, mostly male, young, healthy and not too fat.

13.6 The concept of renal clearance

See Brändström *et al.* (1998) for a comparison of GFR measurements made using iohexol and ^{51}Cr-EDTA.

14.4 Osmotic pressure and colloid osmotic pressure (oncotic pressure)

The relation between protein concentration and the colloid osmotic pressure of blood plasma is complicated and nonlinear (Burton, 1988). One complication is that part of the colloid osmotic pressure is due to the net charge on the proteins, this being pH-sensitive and usually negative. This net charge causes an uneven distribution of inorganic ions across capillary walls (or the artificial membranes of colloid osmometers) and therefore a small difference in their contribution to total osmotic pressure. For calculations relating to that point, see Burton (2000).

The word 'oncotic', which is often used, relates to the swelling of tissues and therefore fails to convey the full significance of colloid osmotic pressure.

Problem 14.19

The latent heat of evaporation depends somewhat on temperature, being less at 100 °C. The value given applies at 30 °C. It is little affected by dissolved solids, so that the concentration of the sweat by evaporation does not need to be considered (Wenger, 1972).

21 pH – two definitions and a possible dilemma for teachers

I looked at 20 textbooks of general chemistry and four of general biochemistry, all published in the period 1981–2002 (Burton, 2007a). All define pH in terms of Eq. (21.1) and the molar concentration of hydrogen or hydronium ions. Six books rightly attribute their definition to Sørensen, who published it in 1909 (with two of the books spelling his name correctly). Unusually, he published his paper in both German and French. (Incidentally, it is often stated that the p in pH, originally written as p_H, stands for 'Potenz', 'puissance' and, in English, 'power'. That may be a good mnemonic point, but was probably not the original intention.) Two of the textbooks note the need to discard the units (mol/L) before taking logarithms (in one case achieving this by dividing by 'the standard concentration' of 1 mol/L). This point is discussed in Chapter 18. One of the books notes that the definition should strictly refer, not to concentration as such, but, more satisfactorily, to activity.

One is more likely to find a full and proper definition of pH in a textbook of physical chemistry, but full treatments, involving ionic activities and electrode potentials (e.g. Bates, 1973) are unsuited to elementary teaching. The pH scales now in general use are those defined by the National Bureau of Standards and the British Standards Institution (Buck *et al.*, 2002).

22.5 The dependence of [AB] on [A] when ([B] + [AB]) is constant

For the general usefulness of Eq. (22.7) in biology see Burton (1998).

Problem 22.11

The DMO method was first used by Waddell and Butler (1959). The DMO is usually ^{14}C-labelled for easier measurement. A complication omitted from the calculation is that allowance must be made for interstitial fluid in tissue samples and this has to be measured too (see Section 6.2).

23.1 Non-bicarbonate buffering

For the actual variation in the buffer value of plasma proteins with pH, see Siggaard-Andersen (1974), a useful source on acid–base balance.

23.3 Bicarbonate buffering

Confusingly, some people refer to dissolved carbon dioxide as carbonic acid and write Eq. (23.4) with $[H_2CO_3]$ in the bottom line. As pointed out, the two concentrations are hugely different. The value of pK_1' shown in Eq. 23.6 applies in human blood at 37 °C, but not under all other conditions of temperature and salt concentration. If you try to find values appropriate in other situations, you may open up quite a can of worms (Burton, 1987; Putnam and Roos, 1991).

For the slowness of CO_2-$H_2CO_3^-$ equilibration in fish blood, see Gilmour (1998).

23.4 $CO_2/HCO_3{}^-$ and non-bicarbonate buffers together

The mathematics of buffering and the relationships amongst pH, $[HCO_3{}^-]$ and PCO_2 in buffer solutions are treated more fully by Burton (1973). (One may simplify that analysis by ignoring those terms that relate to the minor contributions of carbonate ions.)

There has been much controversy regarding the quantitative treatment of acid–base balance (e.g. Siggaard-Andersen and Fogh-Andersen, 1995). Apart from the physiological mechanisms involved, the important points to bear in mind are the principle of electroneutrality and Eq. (23.5).

23.5 The whole body: diet and the titratable acidity of urine

Of at least six definitions and measurement procedures that are used for titratable acidity, the soundest involves titrating samples of urine with NaOH to the pH of plasma at zero PCO_2 (Burton, 1980). If necessary, a known amount of acid is added first to decompose any bicarbonate present. Then carbon dioxide is expelled and sediments are dissolved. The titratable acidity is calculated from the difference between the added amounts of NaOH and acid. With the urine's content of ammonium ion and the content of organic anion subtracted, the titratable acidity is then compatible with the 'ash titratable acidity' of the diet (with allowance for faecal losses). This measure is obtainable by ashing a sample of food resembling that actually eaten and titrating the resulting ash. Alternatively, it may be calculated from the elemental content of the ash as the difference, in equivalents, between the anion-forming elements chlorine, sulphur and phosphorus and the cation-forming elements sodium, potassium, calcium and magnesium.

23.6 Other aspects of acid–base balance

The three principle acid–base variables of pH, PCO_2 and $[HCO_3^-]$ may be graphed in various ways. Because it is impractical to draw three-dimensional graphs, one must decide which two of the variables to choose as axes. To the above three possibilities may be added $\{H^+\}$, $\log PCO_2$ and $\log[HCO_3^-]$ and several combinations have been advocated. My preference in the context of teaching and learning is to plot $[HCO_3^-]$ (y–axis) against PCO_2 (x-axis) with straight lines of equal pH drawn through the origin (Burton, 2000). In the Siggaard-Andersen curve nomogram (Siggaard-Andersen, 1974), $\log PCO_2$ (y-axis) is plotted against pH (x-axis). Another favoured graph is of plasma bicarbonate (y-axis) plotted against pH (x-axis). In this graph changes caused by varying PCO_2 *in vitro* follow a line of negative gradient equal to -1 times the whole-blood buffer value. It can be a useful exercise to compare details of the different kinds of acid–base graph.

Appendix D
Solutions to problems

2 Units: length, area, volume, mass, moles and equivalents

2.1 1 mequiv is 40 mg/2 = 20 mg. The concentration is 1 mmol/L or 2 mequiv/L.

2.2 70 kg $\times \dfrac{1\ \text{lb}}{0.4536\ \text{kg}}$ = 154 lb (154.3 lb would be over-precise compared with '70').

2.3 40 in $\times \dfrac{25.40\ \text{mm}}{1\ \text{in}} \times \dfrac{1\ \text{m}}{1000\ \text{mm}}$ = 1.016 m. The error is 1.6%.

2.4 6 ft $\times \dfrac{0.305\ \text{m}}{1\ \text{ft}}$ = 1.83 m.

2.5 2000 Å $\times \dfrac{0.1\ \text{nm}}{1\ \text{Å}} \times \dfrac{1\ \mu\text{m}}{1000\ \text{nm}}$ = 0.2 μm.

2.6 200 nm = 0.2 μm. (0.2 + 1 + 3)/3 μm = 1.4 μm.

2.7 1 mm = 1000 μm. $(1000\ \mu\text{m})^3 = 10^9\ \mu\text{m}^3$.

2.8 $(12\ \text{in})^2 = (1\ \text{ft})^2$. Therefore 144 in^2 = 1 ft^2.

2.9 (a) 3 ft $\times$ 2 ft = 6 ft^2.   6 ft$^2 \times$ (0.305 m/ft)^2 = 0.56 m^2.

   (b) 0.56 m$^2 \times$ 0.2 m $\times \dfrac{1000\ \text{L}}{\text{m}^3}$ = 112 L.

2.10 Magnesium is divalent. Therefore 2 mequiv Mg = 1 mmol Mg.

Biomedical Calculations   Richard F. Burton
© 2008 John Wiley & Sons, Ltd

2.11 $2000 \text{ mmol} \times \dfrac{1 \text{ mol}}{1000 \text{ mmol}} \times \dfrac{23 \text{ g}}{1 \text{ mol}} = 46 \text{ g.}$

2.12 $400 \text{ mmol} \times \dfrac{1 \text{ mol}}{1000 \text{ mmol}} \times \dfrac{60 \text{ g}}{1 \text{ mol}} = 24 \text{ g.}$

3 Percentages

3.1 $100 \times 15/(60 + 15) = 20\%. \ 100 \times 90/(60 + 90) = 60\%.$ No: the ratio is 3.

3.2   Divide top and bottom of the right-hand expression by N. Then:

$$\text{percentage} = 100[F/N]/(1 + [F/N]).$$

3.3 $F/N = 0.3/(100 - 0.3) = 0.00301.$ The ratio is $0.0417/0.00301 = 13.9.$
This hardly differs from the ratio of percentages because the numbers are on the nearly linear part of the curve in Figure 3.2.

3.4 $F/N = 53/(100 - 53) = 1.13.$ The ratio is $1.13/0.0417 = 27.1.$ This is much greater than the ratio of percentages.

3.5 0.0417 and $13.25.$

3.6   If $F/N = 0.05$, $N = 20F$ and $(N + F) = 21F.$ So $F/(N + F) = 1/21 = 0.0476.$

$$0.0476 \times 500 \text{ g} = 23.8 \text{ g.}$$

Alternatively, as discussed in Chapter 6, one may let the answer $= x \text{ g.}$ Then $F/N = 0.05 = x/(500 - x).$ Then $x = 25 - 0.05x. \ 1.05x = 25. \ x = 25/1.05 = 23.8.$
So the answer is 'no'.

3.7   Calculate heat losses as percentages of 70 and 700 kilocalories per hour.
Radiation: at rest 45.5, in exercise 105. (kilocalories per hour)
Convection: at rest 10.5, in exercise 105.
Evaporation: at rest 14, in exercise 490.

3.8 $100 \times (1300 \text{ g} \times 1 \text{ kg}/1000 \text{ g})/70 \text{ kg} = 1.86\%.$

3.9 $65 \text{ g}/1100 \text{ ml} \times 100 = 5.9\%$ (w/v).

3.10 $70 \text{ kg} \times [100 - (62 + 15)]/100 = 16.1 \text{ kg}$.

3.11   (a) Ignore the 70 kg. $100 \times 62/(100 - 15) = 72.9\%$.
(b) Original (and final) mass of water $= 62\%$ of $70 \text{ kg} = 43.4 \text{ kg}$.
Final percentage of water $= 100 \times 43.4 \text{ kg water}/(70 + 8) \text{ kg total} = 55.6\%$.

3.12   Percentage $= 63\% \times 4/100 = 2.52\%$.
Leucocytes per microlitre $= 2.52/100 \times 7400 = 186$.

4 Composite units I – density

4.1 g mL^{-1} and kg L^{-1}.

4.2   The volume of protein in 100 mL of plasma is $0.73\,P$ mL. The volume of the
rest is $(100 - 0.73\,P)$ mL and its mass is $(100 - 0.73\,P) \text{ mL} \times 1.007 \text{ g mL}^{-1} =$
$1.007(100 - 0.73\,P)$ g. The total mass of 100 mL equals that plus P g, which
is $1.007(100 + 0.27\,P)$ g. Therefore

$$D = \frac{(100.7 + 0.27P) \text{ g}}{100 \text{ mL}}.$$

By rearrangement of this, $P = 370(D - 1.007)$, which is the required formula (see Notes).

4.3   They are all the same.

4.4   Percentage body fat $= (495 \text{ g/cm}^3)/(1.05 \text{ g/cm}^3) - 450 = 21.4\%$.

4.5   2% NaCl (w/w) $= 20 \text{ g NaCl (kg solution)}^{-1}$. d_4^{20} is numerically the same
as the density (kg L^{-1}). The two quantities can be written as

$$\frac{20 \text{ g NaCl}}{1 \text{ kg solution}} \quad \text{and} \quad \frac{1.0125 \text{ kg solution}}{1 \text{ L solution}}.$$

Only by multiplying one by the other (cancelling 'kg solution') do we get an
answer in the right units. It is $20.25 \text{ g NaCl L}^{-1}$.

5 Composite units II – concentration

5.1   The concentration in mmol/kg water. The protein and lipid contribute to
the volume, but not to the mass of water.

5.2 The volume of a cell is due to much more than just water, but it is the concentration in water (molality) that generally determines the physical effects of a solute. It is presumably also the molality that is actually regulated by the cell.

5.3 From Eq. (5.4) the volume is 4 mg ÷ 500 mg L^{-1} = 0.008 L = 8 mL. The two quantities cannot be added or subtracted and division, producing 1/8 mL^{-1}, has no obvious significance. There is therefore only the one way of combining them.

5.4 For calcium, 2 mequiv/L and 1 mEq/L are not the same as 2 mmol/L.

5.5 From Eq. (5.3) the volume is 500 mg ÷ 8.6 mg/L water = 58 L.

5.6 Assume that the 'concentration' of rings is 10 rings/100 geese. Number of rings in population = 30. Population size (equivalent to volume) = 30 rings ÷ 10 rings/100 geese = 300 geese. The relevant equation is (5.4).

6 Aspects of problem solving

6.1 Density = 1.03 kg/L = 1.03 g/cm^3.
Percentage body fat = $(495\ g/cm^3)/(1.03\ g/cm^3) - 450 = 31\%$.

6.2 Work through the original calculation, but substituting 0.2 kg water in the first line. The answer is 1.0 mmol/kg cell water.

6.3 Let the depth of fluid be x cm. Volume = $\pi \times (4.4\ cm)^2 \times x$ cm = 30 cm^3. $x = 30\ cm^3/[\pi \times (4.4\ cm)^2] = 0.49$ cm. The depth of the dish is irrelevant.

6.4 If x mM is the answer, x mM + (x mM)/4 = 1.2 mM. Therefore, $x(1 + 0.25) = 1.2$ and $x = 1.2/(1 + 0.25) = 0.96$. Alternatively, $x/(1.2 - x) = 4$. $x = 4.8 - 4x$. $5x = 4.8$. $x = 4.8/5 = 0.96$.

6.5 Let x be the number of mL required. Then

$$15\ g/L = 50\ g/L \times \frac{x}{20 + x}.$$

Therefore

$$\frac{x}{20 + x} = \frac{15\ g/L}{50\ g/L} = 0.3.$$

Multiplying both sides by $(20 + x)$ gives: $x = 0.3(20 + x) = 6 + 0.3x$. Therefore $x(1 - 0.3) = 6$ and $x = 6/0.7 = 8.6$.

6.6   Period $=$

$$2\,s = 2\pi\sqrt{\frac{l}{g}} = 2\pi\sqrt{\frac{l}{9.8\text{ m s}^{-2}}}.$$

Divide both sides by 2π:

$$1/\pi\,s = \sqrt{\frac{l}{9.8\text{ m s}^{-2}}}.$$

Square both sides:

$$1/\pi^2 s^2 = \frac{l}{9.8\text{ m s}^{-2}}.$$

Therefore $l = 1/\pi^2\,s^2 \times 9.8\text{ m s}^{-2} = 0.99$ m. (So how might you measure out 1 m of string using just your watch?)

7 Making up and diluting solutions

7.1   246.5 g would be needed in making up 1 L, so one quarter of that (61.6 g) is needed in the 250 mL flask. The other information is not required.

7.2

$$\frac{?\text{ g}}{1\text{ L saline}} = \frac{50\text{ g}}{1\text{ L solution}} \times \frac{0.3\text{ mL solution}}{30\text{ mL saline}} = \frac{0.5\text{ g}}{1\text{ L saline}} = 0.5\,\text{g/L}.$$

The factor for converting between mL and L could be included, but twice.

7.3

$$2.5 \times 10^{-6}\text{ M} \times \frac{x\text{ mL}}{25\text{ mL}} = 3 \times 10^{-8}\text{ M}.$$

$$x\text{ mL} = 25\text{ mL} \times \frac{3\times 10^{-8}\text{ M}}{2.5\times 10^{-6}\text{ M}} = 30 \times 10^{-2}\text{ mL} = 0.3\text{ mL}.$$

There is generally no need to allow for the slight increase in total volume which, in this case, is only from 25 to 25.3 mL.

7.4 Bicarbonate is univalent, so 1 mequiv = 1 mmol and there is no need to spell out this conversion in the calculation. In this simple situation each 1 mL of stock solution made up to 1 L gives a concentration of 1 mmol/L, so 25 mL are required. The available volume of stock solution and the molecular mass are irrelevant.

7.5 Mass of sucrose needed =

$$\frac{342.3 \text{ mg}}{1 \text{ mmol}} \times \frac{200 \text{ mmol}}{L} \times \frac{1 \text{ L}}{1000 \text{ mL}} \times 100 \text{ mL} = 6846 \text{ mg (or 6.85 g)}.$$

7.6

$$? \text{ mL stock} = 250 \text{ mL saline} \times \frac{1 \text{ L}}{1000 \text{ mL}} \times \frac{5 \text{ mmol}}{1 \text{ L saline}} \times \frac{1 \text{ mol}}{1000 \text{ mmol}}$$

$$\times \frac{180 \text{ g}}{1 \text{ mol}} \times \frac{1 \text{ L stock}}{90 \text{ g}} \times \frac{1000 \text{ mL}}{1 \text{ L}} = 2.5 \text{ mL stock solution}.$$

8 Calculating drug doses

8.1

$$? \text{ mL} = 0.5 \text{ g} \times \frac{1000 \text{ mg}}{1 \text{ g}} \times \frac{5 \text{ mL}}{250 \text{ mg}} = 10 \text{ mL}.$$

8.2 British:

$$200 \text{ minims} \times \frac{1 \text{ mL}}{16.9 \text{ minims}} = 11.8 \text{ mL}.$$

US:

$$200 \text{ minims} \times \frac{1 \text{ mL}}{16.2 \text{ minims}} = 12.3 \text{ mL}.$$

(12 mL is close enough in both cases.)

8.3

$$0.15 \text{ grain} \times \frac{64.8 \text{ mg}}{1 \text{ grain}} \times \frac{1 \text{ tablet}}{10 \text{ mg}} = 1.0 \text{ tablet.}$$

8.4 $300 \text{ mg/m}^2 \times 1.69 \text{ m}^2 = 507 \text{ mg.}$

8.5

$$\frac{? \text{ mL}}{\text{min}} = \frac{50 \text{ mL}}{50 \text{ mg}} \times \frac{1 \text{ mg}}{1000 \text{ μg}} \times \frac{3 \text{ μg}}{\text{kg min}} \times 68 \text{ kg} = 0.204 \text{ mL/min.}$$

8.6

$$? \text{ mL} = \frac{5 \text{ mL}}{50 \text{ mg}} \times \frac{4 \text{ mg}}{\text{kg}} \times 24 \text{ kg} = 9.6 \text{ mL.}$$

8.7

$$\frac{? \text{ mL}}{\text{dose}} = \frac{5 \text{ mL}}{250 \text{ mg}} \times \frac{30 \text{ mg}}{\text{kg day}} \times \frac{1 \text{ kg}}{2.20 \text{ lb}} \times 150 \text{ lb} \times \frac{1 \text{ day}}{3 \text{ doses}} = \frac{13.6 \text{ mL}}{\text{dose}}.$$

9 More about solutions – electroneutrality, osmotic pressure and activity

9.1 25 mM $NaHCO_3$, 120 mM NaCl, 5 mM KCl, 2 mM $CaCl_2$ (as well as 1 mM $MgSO_4$). Alternatively, you could use 5 mM $KHCO_3$, with 20 mM $NaHCO_3$ and 125 mM NaCl.

9.2 Estimated osmolality $= (2[Na] + [glucose] + [urea]) = (284 + 4.9 + 6.6)$ mmol/L $= 295.5$ mmol/L.

Osmolar gap $= (324 - 295.5) = 28.5$ mosmol/kg water. This is close to the concentration of lithium plus that of the extra anion required to balance its charge ($= 2 \times 14.5 = 29$ mmol/L).

9.3 Osmosity gap $= 2 \times 500 - 149 - 40 - 46 = 765$ mmol/kg water.

$$\text{Anion gap} = 40 + 149 - 46 = 143 \text{ mequiv/kg water.}$$

Here a much larger part of the osmotic pressure is unexplained, but the difference between the two anion gaps is nothing like as big. This is compatible with the presence of high concentrations of substances without net charge.

9.4 285 mosmol/kg water $\times$ 42 L/43 L = 278.4 mosmol/kg water.

9.5 The increase in extracellular water from 14 L to 14.7 L would reduce the sodium concentration in the ratio 14 L/14.7 L = 0.95. For example it might fall from 150 to 143 mmol/kg water.

10 Graphs, straight lines and equations

10.1 $m = 3/2 = 1.5$ and $120/96 = 1.25$
The first estimate is less accurate, because the points are closer together.

10.2 (a) 150 mg $= 9$ mg/g $\times$ 40 g $+ c = 360$ mg $+ c$. Therefore $c = 150$ mg $-$ 360 mg $= -210$ mg.
(b) $y = 9$ mg/g $\times x$ g $- 210$ mg $= 0$, so $x = 210$ mg/(9 mg/g) $= 23.3$ g.

10.3 The coordinates are (0, 2.0) and (4, 5.2).

10.4 $y = 0.417x + 1.25$.

10.5 (a) Gradient $= (100 - 0)\,^{\circ}C/(212 - 32)\,^{\circ}F = 0.556\,^{\circ}C/^{\circ}F$.
$0\,^{\circ}C = 0.556\,^{\circ}C/^{\circ}F \times 32\,^{\circ}F + c$. Therefore $c = -17.8\,^{\circ}C$.
(b) $0.556\,^{\circ}C/^{\circ}F \times 98.6\,^{\circ}F - 17.8\,^{\circ}C = 37.0\,^{\circ}C$.
(c) $x\,^{\circ}C = 0.556\,^{\circ}C/^{\circ}F \times x\,^{\circ}F - 17.8\,^{\circ}C$. $x = 40.1$.
(d) $1/(0.556\,^{\circ}C/^{\circ}F) = 1.80\,^{\circ}F/^{\circ}C$.
(e) $0.556 = 5/9$. $1.80 = 9/5$.

11 On shapes and sizes

11.1 Volume $= {}^4/_3\,\pi \times$ radius$^3 = {}^4/_3\,\pi \times (25/2$ mm$)^3 = 8181$ mm^3, or 8.2 cm^3.

11.2 Volume $= \pi\,(0.4/2\,\mu m)^2 \times 4\,\mu m = 0.50\,\mu m^3$.

11.3 Call all three semi-axes r. Then the expression becomes:

$$4\pi\,[(r^P r^P + r^P r^P + r^P r^P)/3]^{1/P} = 4\pi\,[(r^P r^P)]^{1/P} = 4\pi\,r^2.$$

11.4 1 mm$^3 \times 44/100 \div (5.4 \times 10^6) \times 10^9 \mu m^3/mm^3 = 81.5\,\mu m^3 = 81.5$ fL.

11.5 Volume 88 $\mu m^3 = 51.5 \mu m^2 \times$ thickness, so thickness $= 88 \mu m^3/(51.5 \mu m^2) = 1.71 \mu m$.

11.6 25.4 $\mu m \times 1.71 \mu m = 43 \mu m^2$. The total surface area of the erythrocyte is thus $103 \mu m^2 + 43 \mu m^2 = 146 \mu m^2$.

11.7 Surface area $= 146 \mu m^2 = 4\pi$ (radius of sphere)^2. Therefore the radius of the sphere $=$

$$\sqrt{\frac{146\ \mu m^2}{4\pi}} = 3.41 \mu m.$$

The volume $= \frac{4}{3} \cdot \pi$ (radius)$^3 = \frac{4}{3} \cdot \pi$ $(3.41\ \mu m)^3 = 166\ \mu m^3$.

11.8 The ratio is 166 $\mu m^3/88\ \mu m^3 = 1.89$.

11.9 (0.9% NaCl $\times$ 53 $\mu m^3)/(131\ \mu m^3) = 0.36\%$ NaCl.

11.10 Volume $= 88\ \mu m^3 = \frac{4}{3} \pi$ (radius)^3. Therefore,

$$\text{radius} = \sqrt[3]{\frac{88\ \mu m^3}{1.33\pi}} = 2.76\ \mu m.$$

The diameter is twice that, i.e. 5.5 μm. Surface area $= 4\pi$ (radius)$^2 = 4\pi$ $(2.76\ \mu m)^2 = 95.7\ \mu m^2$. As expected, this is less than the value of 146 μm^2 for a normal-shaped erythrocyte.

11.11 The minimum diameter is 5.5 μm. With those values inserted, the left side of Eq. (11.4) equals zero.

11.12 $m^3 = kg \div kg/m^3 = m^3$.

11.13 The gradient $=$

$$k_1 = \frac{\text{difference between two volumes}}{\text{difference between two heights}} = \frac{64 - 0}{400 - 0} \text{ or } \frac{160 - 64}{1000 - 400} = 0.16.$$

Because $K^2 = 0.16$, $K = 0.4$. So, for this particular cuboid, the ratio of width to height is 0.4.

11.14 $\alpha = 2\sqrt{\pi}$, $\beta, = 2$.

11.15 For $\alpha' = \alpha/\sqrt{\rho}$, $(g/cm^3)^{-0.5} = cm^{1.5}\ g^{-0.5}$. For $\beta' = \beta/\rho$, $(g/cm^3)^{-1} = cm^3\ g^{-1}$.

11.16 The temperature difference would be 10 times greater in the larger mammal. Its internal temperature is therefore an improbable 217 °C.

11.17 $A = K \times W^{(2-b)/3} \times H^b$.

11.18 Yes. $3 \times 0.5 + 0.5 = 2$.

11.19 (a) Equation (11.17): 17 650 cm^2; Eq. (11.22): 17 654 cm^2. (b) Equation (11.17): 28 273 cm^2; Eq. (11.22): 30 732 cm^2.
Thus there is a marked discrepancy with the second (obese) individual.

11.20 $W^{(0.5-0.425)} = 0.431 \times H^{(0.725-0.5)}$.
$W^{0.075} = 0.431 \times H^{0.225}$.
$W = 0.431^{1/0.075} \times H^{0.225/0.075} = 1.34 \times 10^{-5} \times H^3$.
W is thus proportional to H^3. This follows from Eq. (11.21). If a has values c and d in any two versions of the formula, then $[W^c \times H^{(2-3c)}]$ is proportional to $[W^d \times H^{(2-3d)}]$. Dividing both by H^2 and rearranging we then have $[W/H^3]^c$ proportional to $[W/H^3]^d$.

11.21 If $H = 160$ cm, $W = 1.34 \times 10^{-5} \times H^3 = 54.9$ kg.
Equation (11.17): $A = 71.84 \times 54.9^{0.425} \times 160^{0.725} = 15\,620$ cm^2.
Equation (11.22): $A = 1000/6 \times (54.9 \times 160)^{0.5} = 15\,620$ cm^2.

11.22 The point is indeed well above the upper line.

11.23 Let r = radius in mm. Volume = 1 ml = 1000 mm^3 = 60 mm $\times \pi r^2$ mm^2 = $188r^2$ mm^3.
　　$r^2 = 1000/188 = 5.32$ mm^2. $r = \sqrt{5.32} = 2.3$ mm. So the diameter is 4.6 mm.

11.24 1 mm^3 $\times 40/100 \div (4.6 \times 10^6) \times 10^9 \,\mu\text{m}^3/\text{mm}^3 = 87\,\mu\text{m}^3 = 87$ fL.

11.25 0.9% NaCl, if w/v, equals 9 g/L solution. 9 g/L $\times$ (1 mol)/(58.5 g) $\times$ 1000 mmol/mol = 154 mmol/L.
　　This is the osmosity. The osmolarity, incidentally, is about double that ($= 154 \times 2 \times 0.928 = 286$ mosmol/L).

11.26 (a) Surface area =

$$74.4\,\mu\text{m}^2 \times \frac{\text{m}^2}{10^{12}\,\mu\text{m}^2} \times \frac{6.6 \times 10^6}{\text{mm}^3} \times \frac{1000\,\text{mm}^3}{\text{mL}} \times 10\,\text{mL} = 4.9\,\text{m}^2.$$

(b) The film, one molecule thick, had twice the area of the erythrocyte membranes. This suggests that the membranes have a double layer of lipid ('lipid bilayer').

11.27 The working depends on your choice of numbers.

11.28 Proceed as in section 11.12.

11.29 (a) The larger one has a radius that is bigger by a factor of 10^3, a surface area that is bigger by a factor of $(10^3)^2 = 10^6$ and a volume that is bigger by a factor of $(10^3)^3 = 10^9$. Its relative surface area is therefore smaller by a factor of $(10^3)^2/(10^3)^3 = 10^{-3}$. (b) 10^3, as for the radius.

11.30 $A = \alpha'(cWW^{1/3})^{1/2} + \beta'W/cW^{1/3} = (\alpha'c^{1/2} + \beta'/c)W^{2/3}$.

11.31 $A = 71.84 \times 80^{0.425} \times 175^{0.725} = 71.84 \times 6.439 \times 42.29 = 19\,561$ cm^2.

11.32 $A = 1000/6 \times (80 \times 175)^{0.5} = 19\,720$ cm^2. The agreement is good in this instance.

11.33 New constant $= 71.84 \times 10^{-4} \times 100^{0.725} = 0.2025$. $A = 0.2025 \times 75^{0.425} \times 1.70^{0.725} = 1.864$ m^2.

11.34 Choose x, a and b because it is particularly easy to obtain c as $-(ax^2 + bx)$. Moreover, if x, a and b are all chosen to be positive numbers, $(b^2 - 4ac)$ will not be negative, so there will be no problem in taking its square root. Negative numbers do not have meaningful square roots.

12 Body size, body build, fatness and muscularity: unit analysis as an aid to discovery

12.1 Plot upper arm muscle area (y-axis) against (fat-free mass)$^{2/3}$ (x-axis).

12.2 $16.1 \times 90^{2/3} = 323$ kg.

12.3 Total vs (weight class)$^{2/3}$ or (total)$^{3/2}$ vs weight class. The data could also be plotted as logarithms (Chapter 18).

12.4 The necessary factor is 1.27. This value relates to the coefficients in Eqs (12.3) and (12.4), being equal to 20.5/16.1.

12.5 This procedure would rely on the exact validity of the $^2/_3$-power relationship, but the heaviest weightlifters could be disadvantaged by their tendency to have lower percentage muscle contents. It would be fairer only to make the calculation for individuals just within a weight class. Non-scientific considerations may also occur to the reader.

12.6 Percentage body fat $= 100K \times \text{STM} \times \dfrac{\text{height}^2}{\text{body mass}}$.

13 Introducing time

13.1

$$\frac{\min}{660 \text{ notes}} \times \frac{60 \text{ s}}{\min} \times \frac{1000 \text{ ms}}{s} = 91 \text{ ms/note.}$$

13.2 $60 \text{ } \mu\text{m s}^{-1} \times 3600 \text{ s h}^{-1} \times \text{cm}/(10\,000 \text{ } \mu\text{m}) = 21.6 \text{ cm h}^{-1}.$

13.3 $2 \times 9.8 \text{ m s}^{-1} = 19.6 \text{ m s}^{-1}.$

13.4

$$\frac{10 \text{ mL oil}}{\text{day}} \times \frac{0.12 \text{ mg vitamin}}{\text{mL oil}} = 1.2 \text{ mg vitamin day}^{-1}.$$

13.5

$$2 \text{ mL plasma} \times \frac{3.4 \text{ mg Na}}{\text{mL plasma}} = 6.8 \text{ mg Na}$$

13.6

$$\frac{180 \text{ L}}{\text{day}} \times \frac{3.5 \text{ g Na}}{\text{L}} = 630 \text{ g Na day}$$

13.7

$$\frac{6 \text{ L blood}}{\min} \times \frac{150 \text{ mL O}_2}{\text{L blood}} = 900 \text{ mL O}_2/\min$$

13.8

$$\frac{5 \text{ L air}}{\min} \times \frac{4 \text{ L CO}_2}{100 \text{ L air}} = 0.2 \text{ L CO}_2/\min$$

13.9   Question 13.5.

13.10   Volume per unit time × amount per unit volume = amount per unit time.

13.11 $V = 57 \text{ mL}/10 \text{ min} = 5.7 \text{ mL/min.}$

$$\text{GFR} = UV/P = \frac{(2235 \text{ mg}/100 \text{ mL})(5.7 \text{ mL/min})}{122 \text{ mg}/100 \text{ mL}} = 104 \text{ mL/min.}$$

13.12 $V = 85$ mL/10 min $= 8.5$ mL/min.

$$\text{ERPF} = UV/P = \frac{(49.6 \text{ mg}/100 \text{ mL})(8.5 \text{ mL/min})}{0.64 \text{ mg}/100 \text{ mL}} = 659 \text{ mL/min}.$$

13.13   500 mL/min $\times$ 100/(100 − 42) = 862 mL/min.

13.14   1/(1 − 0.20) = 1.25. This leads to a similar rise in colloid osmotic pressure.

13.15

$$\frac{(140 - 24) \times 75.1}{72 \times 1.1} = 110 \text{ ml/min}$$

13.16   Rate of nitrogen excretion $=$ 1200 mmol N/1170 mL $\times$ 1000 mL/L $=$ 1026 mmol N/L. Rate of sulphate excretion $=$ 36.5 mmol SO_4/1170 mL $\times$ 1000 mL/L $=$ 31 mmol SO_4/L.

13.17   385 mL/day = 0.385 L/day. Rate of nitrogen excretion = 668 mmol/L $\times$ 0.385 L/day = 257 mmol/day. Rate of sulphate excretion = 15.2 mmol/L $\times$ 0.385 L/day = 5.8 mmol/day. Compare these rates with those in the previous question.

13.18

$$\frac{211 \text{ mL O}_2 \text{ min}^{-1}}{(193 - 147) \text{ mL O}_2 \text{ (L blood)}^{-1}} = 4.6 \text{ L min}^{-1}.$$

13.19

$$\frac{203 \text{ mL CO}_2 \text{ min}^{-1}}{(525 - 485) \text{ mL CO}_2 \text{ (L blood)}^{-1}} = 5.1 \text{ L min}^{-1}.$$

13.20   (a)

$$\frac{(140 - 75) \times 72}{72 \times 1.8} = 36.1 \text{ mL/min}.$$

(b) 36.1 mL/min $\times$ 1.8 mg/100 mL = 0.65 mg/min.

14 Force, pressure, energy, work and power

14.1 1 kg corresponds to 9.8 N. Therefore 1 N corresponds to $1/9.8 \times 1000$ g/1 kg $= 102$ g.

14.2 1 atm $= 760$ mmHg $= 760$ mmHg $\times$ (1.36 cm H_2O)/(1 mmHg) $= 1034$ cm H_2O.

14.3 1034 cm H_2O corresponds to a depth of about 10 m.

14.4

$$120 \text{ cm blood} \times \frac{10 \text{ mm}}{1 \text{ cm}} \times \frac{1.06 \text{ kg/L}}{13.6 \text{ kg/L}} = 93.5 \text{ mmHg}.$$

14.5 $290 \times 19.3 = 5597$ mmHg. $290 \times 2.57 = 745$ kPa.

14.6 1 mmol/kg water $\times$ 1 mol/1000 mmol $\times$ 68 000 g/mol $= 68$ g/kg water.

14.7 (2065 kcal/day)/(3.7 kcal/g) $= 558$ g.

14.8 120 mmHg $\times$ 0.1333 kPa/(1 mmHg) $= 16.0$ kPa. This was chosen as being a typical healthy systolic blood pressure.

14.9 The calorie is conveniently defined in terms of the warming of water.

14.10 4.2 differs from 4.186 by only 0.3 per cent; few measurements would be anything like that accurate.

14.11 11 000 kJ/day $\div$ 4.186 kJ/kcal $= 2628$ kcal/day.

14.12 43 g of sugar $\times$ 3.7 kcal/g $+$ 24.6 g of other carbohydrate $\times$ 4.2 kcal/g $+$ 3.9 g of protein $\times$ 4.4 kcal/g $+$ 15.9 g of fat $\times$ 9.3 kcal/g $= 427.5$ kcal.

14.13 1700 kcal/day $\times$ 4.186 kJ/kcal $= 7116$ kJ/day.

14.14 $(2600 - 2500)$ kcal/day $\times$ 365 day $= 36\ 500$ kcal.

$$36\ 500 \text{ kcal} \div 9.3 \text{ kcal/g} = 3925 \text{ g (3.9 kg)}.$$

14.15 (1700 kcal)/(360 L O_2) $= 4.7$ kcal/L O_2. This corresponds to the metabolism of fat, a combination of protein and carbohydrate, or a combination of all three.

14.16 1 kilowatt hour $= 1000$ J/s $\times$ 3600 s $= 3.6 \times 10^6$ J.

14.17 $m^3 \times N/m^2 = N$ m.

14.18 $1 L = 10^{-3} \text{ m}^3$. $1 \text{ atm} = 101.3 \text{ kN m}^{-2}$.

$$1 \text{ L atm} = 10^{-3} \text{ m}^3 \times 101.3 \text{ kN m}^{-2} = 0.1013 \text{ kN m} = 101.3 \text{ J}.$$

14.19 (a) $100 \text{ J} \times 1 \text{ kJ}/1000 \text{ J} \div 2.43 \text{ kJ/g} = 0.041 \text{ g}$. (b) $100 \text{ W} = 100 \text{ J/s} \cdot 0.041$ g/s $\times$ 3600 s/h $= 148$ g/h. The evaporative water loss in the absence of sweating (i.e. the 'insensible water loss') is typically about 50 mL/h.

14.20 (a) $38 \text{ mol} \times 12 \text{ kcal/mol} = 456 \text{ kcal}$. (b) $(456 \text{ kcal})/(666 \text{ kcal}) \times 100 = 68\%$.

15 Lessons from another formula

15.1 $(\text{kg m s}^{-2}) \times \text{s} \times \text{m}^{-2} = \text{kg m}^{-1} \text{s}^{-1}$.

15.2 Only 'd' is false. It is true that flow rate decreases with increasing P_2.

15.3

$$\frac{1600 \text{ dyne s}}{\text{cm}^5} \times \frac{1 \text{ N}}{100\,000 \text{ dyne}} \times \frac{\text{min}}{60 \text{ s}} \times \left(\frac{100 \text{ cm}}{\text{m}}\right)^5 \times \frac{1000 \text{ cm}^3}{\text{L}}$$

$$\times \frac{(0.01 \text{ m})^3}{\text{cm}^3} \times \frac{\text{mmHg m}^2}{133.3 \text{ N}} = 20 \text{ mmHg min L}^{-1}.$$

16 Heat and temperature

16.1 The exponent equals $10/10 = 1$. So $Q_{10} = R_2/R_1$.

16.2 $37.0 \,^\circ$C is 310.15 K. $100 \times 0.1/310.15 = 0.03\%$.

16.3 (a) $80 \text{ kg} \times 0.8 \text{ kcal}/(\text{kg} \,^\circ\text{C}) \times 1 \,^\circ\text{C} = 64 \text{ kcal}$.

(b) $\dfrac{128 \text{ kcal}}{\text{h}} \times \dfrac{\text{kg} \,^\circ\text{C}}{0.8 \text{ kcal}} \times \dfrac{1}{80 \text{ kg}} = 2 \,^\circ\text{C per hour.}$

16.4 Let final temperature $= t$. The increased heat content of the water (in cal) equals the fall in heat content of the body. The former is: $650 \text{ g} \times 1.0 \text{ cal/g}$ $^\circ\text{C} \times (t - 20 \,^\circ\text{C})$. The latter is $65\,000 \text{ g} \times 0.8 \text{ cal/g} \,^\circ\text{C} \times (37 \,^\circ\text{C} - t)$. Equating these, and simplifying the resulting equation gives: $(650t - 13\,000) \text{ cal} = (1\,924\,000 - 52\,000t) \text{ cal}$. Thus $52\,650t = 1\,937\,000$ and $t = 36.8 \,^\circ\text{C}$. This is a fall of $0.2 \,^\circ\text{C}$.

16.5 Q_{10} for 25–35 °C = 7.0/3.6 = 1.9. Q_{10} for 10–20 °C = 2.4/0.75 = 3.2. Q_{10} for 20–25 °C = $(3.6/2.4)^{(10/5)}$ = 2.25.

16.6 $Q_{10} = (273.15 + 30)/(273.15 + 20) = 1.03$.

16.7 (a) 5 L/min × 1000 g/L × 1 cal/(°C g) × 25 °C × 1 min/(60 s) = 2083 cal s^{-1}. (b) 2083 cal s^{-1} × 4.186 J/cal = 8719 J s^{-1} = 8719 W (i.e. 8.7 kW).

17 Gases: dry and wet gas mixtures, partial pressures, gases in solution

17.1

$$V_{STPD} = 15.0 \text{ L} \times \frac{(98.0 \text{ kPa} - 1.7 \text{ kPa})}{101.3 \text{ kPa}} \times \frac{273.15 \text{ K}}{288 \text{ K}} = 13.5 \text{ L}.$$

17.2

$$V_{BTPS} = V_{ATPS} \times \frac{(P_{amb} - w)}{(P_{amb} - 47.0)} \times \frac{310}{(273 + T_{amb})}.$$

17.3 $(1 - 1.16 \times 10^{-4} \times 5500 + 4.4 \times 10^{-9} \times 5500^2) = 0.495$.

17.4

$$\frac{0.0821 \text{ L atm}}{\text{mol K}} \times \frac{1}{1.02 \text{ atm}} \times 300 \text{ K} \times 0.2 \text{ mol} = 4.83 \text{ L}.$$

The question must be: what is the volume of helium in the balloon?

17.5 22.4 L is 1 mol and therefore 32 g. Density = (32 g)/(22.4 L) = 1.43 g/L. 1.43 g/L × (1 kg)/(1000 g) × (1000 L)/(1 m^3) = 1.43 kg/m^3.

17.6 78.1/100 × 760 mmHg = 594 mmHg.

17.7 6.3 kPa/100 kPa × 100 = 6.3 %.

17.8 (a) ATPS to STPD:

$$4.00 \text{ L} \times \frac{96 - 2.3}{101.3 - 0} \times \frac{273}{273 + 20} = 3.45 \text{ L}.$$

(b) STPD to BTPS:

$$2.00 \text{ L} \times \frac{760 - 0}{760 - 47} \times \frac{273 + 37}{273 + 0} = 2.42 \text{ L.}$$

(c) ATPS to BTPS:

$$20.0 \text{ L} \times \frac{740 - 12.8}{740 - 47} \times \frac{273 + 37}{273 + 15} = 22.6 \text{ L.}$$

(d) BTPS to STPD:

$$20.0 \text{ L} \times \frac{752 - 47}{760 - 0} \times \frac{273}{273 + 37} = 16.3 \text{ L.}$$

(e) BTPS to non-standard condition:

$$3.00 \text{ L} \times \frac{95 - 6.3}{95 - 1.7} \times \frac{273 + 15}{273 + 37} = 2.65 \text{ L.}$$

17.9 (a) $14\,000 \text{ L}/(22.4 \text{ L/mol}) = 625 \text{ mol/day}$. (b) $100 \times 37.7/759 = 4.97\%$. (c) The molar mass of water is 18. $625 \text{ mol/day} \times 4.97/100 \times 18 \text{ g/mol} = 559$ g/day.

A typical rate of water loss in expired air is generally said to be about 400 mL/day. Actual losses depend on metabolic rate and the moisture content of the inspired air.

Small mammals have much higher relative metabolic rates than we do and, for desert rodents, there is particular value in cooling the expired air to conserve water.

17.10 $40 \text{ mmHg} \times 0.03 \text{ mmol}/(\text{L mmHg}) = 1.20 \text{ mmol/L}$.

17.11 $100 \text{ mmHg} \times 0.0013 \text{ mmol}/(\text{L mmHg}) = 0.13 \text{ mmol/L}$.

17.12

$$\frac{0.030 \text{ mmol CO}_2}{\text{L mmHg}} \times \frac{22.4 \text{ mL CO}_2}{1 \text{ mmol CO}_2} \times \frac{7.50 \text{ mmHg}}{1 \text{ kPa}} = \frac{5.04 \text{ mL CO}_2}{\text{L} \quad \text{kPa}}$$

$$= 5.04 \text{ mL CO}_2 \text{ L}^{-1} \text{ kPa}^{-1}.$$

18 Introduction to logarithms

18.1   If log $y = 1 \times \log x +$ constant, $y =$ constant $\times x$.

18.2   (a) log (AC/B). (b) log (ABE^D).

18.3   (a) AB/C. (b) D^3.

18.4   1.30103, 3.30103, -1.69897.

18.5 $0.5 = 1/2$, so log $0.5 = -0.301$. Log $5 = 0.699$ (see above), so log $(1/5) = -0.699$. Alternatively, $1/5 = 2/10$, so log $0.2 = \log 2 - \log 10 = -0.699$.

18.6   (a) $0.699 - 0.903 = -0.204$. (b) $0.204 + 0.602 = 0.806$.

18.7 -0.43.

18.8   Log $7 \log^7 \approx \log 50 = \log 100 - \log 2 = 2 - 0.301 = 1.699$. Log $7 \approx 1.699/2 = 0.8495$. The correct value is 0.845.

18.9 $4 \log 3 \approx 3 \log 2 + \log 10 = 1.903$. Log $3 \approx 1.903/4 = 0.476$. The correct value is 0.477.

19 Exponential time courses

19.1   The exponent bt must be free of units. Therefore the unit for b must be min^{-1}.

19.2   The rate of excretion must be proportional to the concentration in the plasma. Therefore the GFR would have to be constant and the plasma concentration would have to be affected by no other process.

19.3   (d) Gradient $= -0.155$ min^{-1}. Rate constant $= -0.155$ min$^{-1}/0.4343 = -0.357$ min^{-1}. (f) -0.36 min^{-1}. (g) -0.357 min^{-1}. $-0.357 \times 0.4343 = -0.155$ min^{-1}. (h) $y = 100e^{-0.357t}$.

20 Nernst equations in physiology and biochemistry: logarithms and 'RT/zF'

20.1   ln $(5/150) = -3.40$.

20.2   A minus sign must precede the right-hand expression. (Recall that log $(1/x) = -\log x$.)

20.3 $61.5 \text{ mV} \times \log(5/150) = -90.8 \text{ mV}.$

20.4 $-61.5 \text{ mV} \times \log(110/5) = -82.6 \text{ mV}.$

20.5 $-61.5 \text{ mV} \times \log(26/13) = -18.5 \text{ mV}.$ This is much less negative than typical membrane potentials, so the internal concentration is higher than would result from simple diffusion equilibrium. This could result from active transport of bicarbonate ions inwards or of hydrogen ions outwards.

20.6

$$E_{Ca} = \frac{61.5}{2} \log \frac{10}{1} = 30.75 \text{ mV}.$$

The membrane could be a calcium-selective membrane as used in a 'calcium electrode'. What is illustrated here is the fact that such an electrode is less sensitive than those used for measuring univalent ions; the voltage change for a given proportionate change in concentration is much less.

21 pH – two definitions and a dilemma for teachers

21.1 $10^{-1.1} = 0.08 \text{ M}.$

21.2 (a) 40 nmol L^{-1}. (b) $(7.4 - 7.1) = 0.3 = \log 2.$ So the answer is 2×40 nmol L^{-1}. (c) No.

21.3 45 and 35 nmol L^{-1}. A personal matter; some people think so.

22 Equilibrium constants, the Henderson–Hasselbalch equation, dose–response curves

22.1 (a) $10^{(7.4-3.5)} = 10^{3.9} = 7943.$ (b) $10^{(7.4-3.3)} = 10^{4.1} = 12\,600.$ (c) Diazepam.

22.2 Because K and $[A]$ can be added, K must be in mmol/L.

22.3 $[s] = K_M.$ This is because

$$\frac{V_{max} K_m}{K_m + K_m} = \frac{V_{max}}{2}.$$

22.4 When converted to logarithms, these values are: $-0.52, 0.0, 0.48, 1.0, 1.48,$ $2.0, 2.48, 3.0, 3.48$ and 4.0.

22.5 Log $([lactate]/[lactic\ acid]) = pH - pK = 7.4 - 4.6 = 2.8$. The ratio is $10^{2.8} = 631$. So there is very little lactic acid in the blood.

22.6 $[base]/[acid] = 10^{(pH-pK)} = 10^{-1} = 0.1$. So $[acid]/[base] = 10$. Percentage charged $= 100 \times 10/(10 + 1) = 91\%$.

22.7 Bupivacaine: $[base]/[acid] = 10^{(7.4-8.1)} = 0.20$. Therefore $[acid] = 5[base]$. $([acid] + [base]) = (5 + 1)[base]$. Percentage as base $= 100[base]/([acid] + [base]) = 17\%$. For lignocaine, calculated in the same way, the percentage is 24%.

22.8 $[HPO_4^{2-}]/[H_2PO_4^-] = 10^{(7.4-6.8)} = 10^{0.6} = 4$. (Alternatively, $\log([HPO_4^{2-}]/[H_2PO_4^-]) = 7.4 - 6.8 = 0.6 = 2\log 2 = \log 4$.)
$\quad\quad [HPO_6^{2-}] + [H_2PO_4^-] = 1.2\ mM = 4 \times [H_2PO_4^-] + [H_2PO_4^-] =$
$\quad\quad 5 \times [H_2PO_4^-]$.
$\quad\quad [H_2PO_4^-] = (1.2\ mM)/5 = 0.24\ mM$.
$\quad\quad [HPO_4^{-2}] = 1.2\ mM - 0.24\ mM = 0.96\ mM$.

22.9 $[PO_4^{3-}]/[HPO_4^{2-}] = 10^{(7.4-12.4)} = 10^{-5} = 0.00001$.
$\quad\quad [H_2PO_4^-]/[H_3PO_4] = 10^{(7.4-2.0)} = 10^{5.4} = 2.5 \times 10^5$.

22.10 Because $[H_2O]$ is not included in the formula for the dissociation constant.

22.11 (a) $[base]/[acid] = 10^{(7.40-6.13)} = 18.62$. (b) $[base] = 18.62[acid]$.
$\quad\quad 1$ mmol/kg water $= (1 + 18.62)[acid]$ mmol/kg water.
$\quad\quad [acid] = 1/19.62 = 0.051$ mmol/kg water.
$\quad\quad$ (c) $[base] = (0.35 - 0.051) = 0.30$ mmol/kg water.
$\quad\quad$ (d) $[base]/[acid] = 0.30/0.051 = 5.89$. This $= 10^{(pH-pK)}$.
$\quad\quad pH - pK = \log 5.89 = 0.77$. $pH = 6.13 + 0.77 = 6.90$.

23 Buffering and acid–base balance

For some of these calculations, one may use either Eq. (23.1) or Eq. (23.2).
23.1 $0.2\ M/0.1\ M = 2 = 10^{(pH-4.8)}$. $\log 2 = 0.3 = pH - 4.8$. $pH = 5.1$.

23.2 $[base]/[acid] = 10^{(7.3-8.3)} = 0.1$.
$\quad\quad$ Percentage dissociation $= 100 \times 0.1/(1 + 0.1) = 9\%$.

23.3 $[HPO_4^{2-}]/[H_2PO_4^-] = 4 = 10^{(7.4-pK)}$.
$\quad\quad \log 4 = 2\log 2 = 0.6 = \log\{10^{(7.4-pK)}\} = 7.4 - pK \cdot pK = 7.4 - 0.6$
$\quad\quad = 6.8$.

23.4   The graph shows that the buffer value decreases from nearly 0.6 (actually 0.576) to nearly 0.2 (actually 0.190) between pH 8.0 and pH 7.0. Therefore about three times as much buffer would be needed.

23.5 $2.303/(2 + 1 + 1) = 0.5758$. This has no unit.

23.6   0.03 mmol/(L mmHg) $\times$ (1 mmHg)/0.1333 kPa $= 0.225$ mmol/kPa. So substitute 0.223 for 0.03.

23.7   Because electroneutrality must be preserved, the ionization state of the HEPES cannot change. Therefore $(pH - pK)$ cannot change either. This means that when the pH falls with rising temperature, so too must the pK (i.e. to 7.31 at 37 °C). It is the change in pK that forces the change in pH.

23.8   (a) From Eq. (23.5), $[B]/[BH^+] = 10^{(7.4-7.55)} = 0.71$. $[B] = 0.71[BH^+]$.
       $(1 + 0.71)[BH^+] = 20$ mM. $[BH^+] = (20$ mM$)/1.71 = 11.7$ mM.
       $[B] = 20$ mM $- 11.7$ mM $= 8.3$ mM.
     (b) Now $[BH^+] = 11.7$ mM $+ 4$ mM $= 15.7$ mM.
       $[B] = 8.3$ mM $- 4$ mM $= 4.3$ mM.
       $pH = 7.55 + \log [B]/[BH^+] = 7.55 + \log (0.274) = 7.55 - 0.56 = 6.99$.
     (c) Take pH 6.99 as pH 7.0, which is 0.4 less than 7.4. The 4 mM of HCl is 4 mequiv L^{-1}. Buffer value $= (4$ mequiv L$^{-1})/0.4 = 10$ mequiv L^{-1} pH^{-1}.
     (d) Buffer value $=$

$$\frac{2.30 \times 20 \text{ mequiv/L}}{2 + 10^{(7.2-7.55)} + 10^{(7.55-7.2)}} = 9.8 \text{ mequiv } L^{-1} pH^{-1}.$$

23.9

$$pH = 6.1 + \log \frac{30 \text{ mmol } L^{-1}}{(0.03 \text{ mmol } L^{-1} \text{ mmHg}^{-1}) \times 35 \text{ mmHg}} = 7.56.$$

23.10

$$7.10 = 6.10 + \log \frac{[HCO_3{}^-]}{0.03 \times 20}. \quad 1.00 = \log \frac{[HCO_3{}^-]}{0.6}$$

$$= \log [HCO_3{}^-] - \log 0.6.$$

$\log [HCO_3{}^-] = 1.00 - 0.22 = 0.78$. $[HCO_3{}^-] = 6$ mmol/L. This is a compensated metabolic acidosis.

23.11

$$7.19 = 6.10 + \log \frac{[HCO_3{}^-]}{0.226 \times 11.5}.$$

$$1.09 = \log \frac{[HCO_3{}^-]}{2.60}$$

$$= \log [HCO_3{}^-] - \log 2.60.$$

$\log [HCO_3{}^-] = 1.09 + 0.41 = 1.50$. $[HCO_3{}^-] = 32$ mmol/L (i.e. somewhat raised in compensation for the respiratory acidosis).

23.12 Recall that $\log 2 = 0.30$ (0.3010). (a) pH $= 7.39 - 0.30 = 7.09$. (b) pH $= 7.39 + 0.30 = 7.69$.

23.13 $\log 30 - \log 22 = 0.135 =$ difference in pH. That this exceeds the normal span of pH (i.e. 0.1) implies that PCO_2 is to some extent adjusted accordingly.

23.14 (a) The potassium ion would need to be balanced by anion. This would be bicarbonate, so the effect would be alkalinizing. (b) The citric acid would turn to carbon dioxide and water, leaving nothing else to affect acid–base balance. (c) The sulphur would be oxidized to sulphate, which would replace some of the bicarbonate in the body.

References

Arnold, G.J. (1998) Refinements in the dimensional analysis method of dose calculation problem-solving. *Nurse Educ.* **23**, 22–26.

Bardeen, C.R. (1920) The height–weight index of build in relation to linear and volumetric proportions and surface-area of the body during post-natal development. Contrib. Embryol. No. 46. *Carnegie Instit. Washington Bull.* **9**, 483–554.

Barlett, H.L., Puhl, S.M., Hodgson, J.L. and Buskirk, E.R. (1991) Fat-free mass in relation to stature: ratios of fat-free mass to height in children, adults, and elderly subjects. *Amer. J. Clin. Nutr.* **53**, 1112–1116.

Bates, R.G. *Determination of pH. Theory and Practice*, 2nd edn. John Wiley and Sons, New York, 1973.

Brändström, E., Grzegorczyk, A., Jacobsson, L., Friberg, P., Lindahl, A. and Aurell, M. (1998) GFR measurement with iohexol and ^{51}Cr-EDTA. A comparison of the two favoured GFR markers in Europe. *Nephrol. Dial. Transplant.* **13**, 1176–1182.

Buck, R.P., Rondinini, S., Covington, A.K., Baucke, F.G.K., Brett, C.M.A., Camões, M.F., Milton, M.J.T., Mussini, T., Naumann, R., Pratt, K.W., Spitzer, P. and Wilson, G.S. (2002) Measurement of pH. Definitions, standards and procedures (IUPAC Recommendations 2002). *Pure Appl. Chem.* **74**, 2169–2200.

Burton, R.F. (1973) The roles of buffers in body fluids: mathematical analysis. *Respir. Physiol.* **18**, 34–42.

Burton, R.F. (1980) Acid and base excretion: assessment and relationships to diet and urine composition. *Comp. Biochem. Physiol.* **66A**, 371–375.

Burton, R.F. (1983a) Inorganic ion pairs in physiology: significance and quantitation. *Comp. Biochem. Physiol.* **74A**, 781–785.

Burton, R.F. (1983b) Cell composition as assessed from osmolality and concentrations of sodium, potassium and chloride: total contributions of other substances to osmolality and charge balance. *Comp. Biochem. Physiol.* **76A**, 161–165.

Burton, R.F. (1983c) The composition of animal cells: solutes contributing to osmotic pressure and charge balance. *Comp. Biochem. Physiol.* **76B**, 663–671.

Burton, R.F. (1987) On calculating concentrations of "HCO_3" from pH and PCO_2. *Comp. Biochem. Physiol.* **87A**, 417–422.

Burton, R.F. (1988) The protein content of extracellular fluids and its relevance to the study of ionic regulation: net charge and colloid osmotic pressure. *Comp. Biochem. Physiol.* **90A**, 11–16.

Burton, R.F. (1998) *Biology by Numbers*. Cambridge University Press, Cambridge.

Burton, R.F. (2000) *Physiology by Numbers*, 2nd edn. Cambridge University Press, Cambridge.

Burton, R.F. (2007a) Defining and teaching pH. *J. Chem. Educ.* **84**, 1129.

Burton, R.F. (2007b) Why is the body mass index calculated as mass/height^2, not as mass/height^3? *Ann. Hum. Biol.* **34**, 656–663.

Burton, R.F. (2008) Estimating body surface area from mass and height: Theory and the formula of Du Bois and Du Bois **35**, 170–184.

Burton, R.F. and Cameron N. (2008) Body fat and skinfold thicknesses: a dimensional analytical approach (in preparation).

Canham, P.B. and Burton, A.C. (1968) Distribution of size and shape in populations of normal human red cells. *Circul. Res.* **22**, 405–422.

Cockcroft, D.W. and Gault, M.H. (1976) Prediction of creatinine clearance from serum creatinine. *Nephron* **16**, 31–41.

Dorwart, W.V. and Chalmers, L. (1975) Comparison of methods for calculating serum osmolality from chemical concentrations, and the prognostic value of such calculations. *Clin. Chem.* **21**, 190–194.

Du Bois, E.F. (1938) The mechanism of heat loss and temperature regulation. *Ann Intern. Med.* **12**, 388–395.

Ebel, H.F., Bliefert, C. and Russey, W.E. (2004) *The Art of Scientific Writing*. Wiley-VCH: Weinheim.

Ellis, K.J., Abrams, S.A. and Wong, W.W. (1999) Monitoring childhood obesity: assessment of the weight/height^2 index. *Am. J. Epidemiol.* **150**, 939–946.

Ford, L.E., Detterline, A.J., Ho, K.K. and Cao, W. (2000) Gender- and height-related limits of muscle strength in world weightlifting champions. *J. Appl. Physiol.* **89**, 1061–1064.

Frisancho, A.R. and Tracer, D.P. (1987) Standards of arm muscle by stature for the assessment of nutritional status of children. *Am. J. Phys. Anthropol.* **73**, 459–465.

Gilmour, K.M. (1998) Causes and consequences of acid–base disequilibria. In S.F. Perry and B.L. Tufts, eds. *Fish Respiration*. Academic Press, San Diego, CA, pp. 321–348.

Glenday, C., ed. (2007) *Guinness World Records 2007*. Guinness World Records Ltd.

Gorter, E. and Grendel, F. (1925) On bimolecular layers of lipoids on the chromocytes of the blood. *J. Exp. Med.* **41**, 439–443.

Jackson, A.S., Stanforth, P.R., Gagnon, J., Rankinen, T., Leon, A.S., Rao, D.C., Skinner, J.S., Bouchard, C. and Wilmore, J.H. (2002) The effect of sex, age and race on estimating percentage body fat from body mass index: The Heritage Family Study. *Int. J. Obes.* **26**, 789–796.

Keys, A. and Brožek, H. (1953) Body fat in adult man. *Physiol. Rev.* **33**, 245–325.

Levine, I.N. (2003) *Physical Chemistry*, 5th edn. McGraw–Hill, New York.

Lietzke, M.H. (1956) Relation between weight-lifting totals and body weight. *Science* **124**, 486–487.

McMahon, T.A., Bonner, J.T. (1983) *On Size and Life*. Scientific American Books, New York.

Prange, H.D., Anderson, J.F. and Rahn, H. (1979) Scaling of skeletal mass to body mass. *Am. Naturalist* **113**, 109–122.

Putnam, R.W. and Roos, A. (1991) Which value for the first dissociation constant of carbonic acid should be used in biological work? *Am. J. Physiol.* 260 (Cell Physiol. 29), C1113–C1116.

Quetelet, A. (1835) *Sur l'homme et le développement de ses facultés ou essai de physique sociale*, Vol. 2. Bachelier, Paris.

Quetelet, A. (1869) *Physique sociale ou essai sur le développement de ses facultés de l'homme*, Vol. 2. C. Muquardt, Brussels.

Ranatunga, K.W. (1984) The force–velocity relation of rat fast- and slow-twitch muscles examined at different temperatures. *J. Physiol.* **351**, 517–529.

Reinhardt, W.H. and Chien, S. (1985) Roles of cell geometry and cellular viscosity in red cell passage through narrow pores. *Am. J. Physiol.* 248 (Cell Physiol. 17), C473–C479.

Rohrer, F. (1908) Eine neue Formel zur Bestimmung der Körperfülle. *Korrespondenze-Blatt der Deutschen Gesellschaft für Anthropologie, Ethnologie und Urgeschichte* **39**, 5–7.

Siggaard-Andersen, O. (1974) *The Acid–Base Status of the Blood*, 4th edn. Munksgaard, Copenhagen.

Siggaard-Andersen, O. and Fogh-Andersen, N. (1995) Base excess or buffer base (strong ion difference) as measure of a non-respiratory acid–base disturbance. *Acta Anaesthesiol. Scand.* **106**, 123–128.

Siri, W.E. (1961) Body composition from fluid spaces and density: analysis of method. In *Techniques for Measuring Body Composition*, eds J. Brozek and A. Henschel. National Academy of Sciences, Washington, DC, pp. 223–244.

Sood, M.M. and Richardson, R. (2007) Negative anion gap and elevated osmolar gap due to lithium overdose. *Can. Med. Assoc. J.* **176**, 921–923.

Sörensen, S.P.L. (1909) Enzymstudien. II. Mitteilung. Über die Messung und die Bedeutung der Wasserstoffionenkonzentration bei enzymatischen Prozessen. *Biochem. Zeitschr.* **21**, 131–304.

Sørensen, S.P.L. (1909) Études enzymatiques. II. Sur la mesure et l'importance da la concentration des ions hydrogène dans les réactions enzymatiques. *Compt. Rend. Trav. Lab. Carlsberg.* **8**, 1–168.

Thomsen, K. (2004) Approximations to the surface area of the scalene ellipsoid, Numericana. www.numericana.com/answer/ellipsoid.htm#thomsen (accessed 10 December 2007).

Van Slyke, D.D., Hiller, A., Phillips, R.A., Hamilton, P.B., Dole, V.P., Archibald, R.M. and Eder, H.A. (1950) The estimation of plasma protein concentration from plasma specific gravity. *J. Biol. Chem.* **183**, 331–347.

Waddell, W.J. and Butler, T.C. (1959) Calculation of intracellular pH from the distribution of 5,5-dimethyl-2,4-oxazolidendione (DMO). Application to skeletal muscle of the dog. *J. Clin. Invest.* **38**, 720–729.

Warton, D.I., Wright, I.J., Falster, D.S., Westoby, M. (2006) Bivariate line-fitting methods for allometry. *Biol. Rev.* **81**, 259–291.

Wenger, C.B. (1972) Heat of evaporation of sweat: thermodynamic considerations. *J. Appl. Physiol.* **32**, 456–459.

West, L.J., Pierce, C.M., Thomas, W.D. (1962) Lysergic acid diethylamide: its effects on a male Asiatic elephant. *Science* **138**, 1100–1103.

White, C.R. and Seymour, R.S. (2003) Mammalian basal metabolic rate is proportional to body mass$^{2/3}$. *Proc. Natl Acad. Sci.* **100**, 4046–4049.

Index